MODEL WATER SHARING AGREEMENTS FOR THE TWENTY-FIRST CENTURY

SPONSORED BY
Environmental and Water Resources Institute of ASCE
and Laws and Institutions Committee of ASCE

EDITED BY
Stephen E. Draper

1801 ALEXANDER BELL DRIVE
RESTON, VIRGINIA 20191–4400

Abstract: When water resources are shared by two or more autonomous political entities, the timing and magnitude of the respective individual uses have been continual sources of conflict. Recognizing the need for clear standards and principles for effective and efficient water sharing, ASCE's Task Committee for the Shared Use of Transboundary Water Resources (SUTWR) reviewed existing transboundary agreements to develop a series of model codes to limit the potential for conflict while providing an appropriate balance between efficient use of the water resource for economic purposes, public health, and ecological protection. All three models—Coordination and Cooperation, Limited Purpose, and Comprehensive Management—focus on the allocation and use of shared waters and on resolving conflicts involving such waters. These three water sharing agreements can be used directly within the United States context and may be used with minor alteration in the international context.

Library of Congress Cataloging-in-Publication Data

Model agreements for the shared use of transboundary water resources / edited by Stephen E. Draper ; sponsored by Environmental and Water Resources Institute of ASCE and Laws and Institutions Committee of ASCE.
 p. cm.
 Includes bibliographical references and index.
 ISBN 0-7844-0614-6
 1. Water—Law and legislation. 2. Water rights (International law) 3. International rivers. 4. Treaties. I. Draper, Stephen E. II. Environmental and Water Resources Institute (U.S.). Laws and Institutions Committee.

K3496 .M63 2002
341.4'42--dc21   2002018528

Any statements expressed in these materials are those of the individual authors and do not necessarily represent the views of ASCE, which takes no responsibility for any statement made herein. No reference made in this publication to any specific method, product, process, or service constitutes or implies an endorsement, recommendation, or warranty thereof by ASCE. The materials are for general information only and do not represent a standard of ASCE, nor are they intended as a reference in purchase specifications, contracts, regulations, statutes, or any other legal document. ASCE makes no representation or warranty of any kind, whether express or implied, concerning the accuracy, completeness, suitability, or utility of any information, apparatus, product, or process discussed in this publication, and assumes no liability therefore. This information should not be used without first securing competent advice with respect to its suitability for any general or specific application. Anyone utilizing this information assumes all liability arising from such use, including but not limited to infringement of any patent or patents.

ASCE and American Society of Civil Engineers—Registered in U.S. Patent and Trademark Office.

Photocopies: Authorization to photocopy material for internal or personal use under circumstances not falling within the fair use provisions of the Copyright Act is granted by ASCE to libraries and other users registered with the Copyright Clearance Center (CCC) Transactional Reporting Service, provided that the base fee of $8.00 per article plus $.50 per page is paid directly to CCC, 222 Rosewood Drive, Danvers, MA 01923. The identification for ASCE Books is 0-7844-0614-6/01/ $8.00 + $.50 per page. Requests for special permission or bulk copying should be addressed to Permissions & Copyright Dept., ASCE.

Copyright © 2002 by the American Society of Civil Engineers.
All Rights Reserved.
Library of Congress Catalog Card No: 2002018528
ISBN 0-7844-0614-6
Manufactured in the United States of America.

PREFACE

This report is the Final Report of the Task Committee for the Shared Use of Transboundary Water Resources (SUTWR). This Report is provided to the Laws & Institutions Committee of the Environmental & Water Resources Institute and American Society of Civil Engineers (ASCE) for publication as a Task Committee Report. Following publication, the SUTWR Task Committee will recommend to its parent committee, the Water Regulatory Standards Committee, that Part I of the Final Report be recommended as a Standard of the American Society of Civil Engineers.

The Water Laws Committee of the Water Resources Planning and Management Division of ASCE of the American Society of Civil Engineers (ASCE) created the Model Water Code Project in 1990 under the director of Professor Ray Jay Davis of the Brigham Young University School of Law. The purpose of the project was to develop statutory provisions intended for adoption by state governments for allocating water rights among competing interests and for resolving quantitative conflicts over water.

After Professor Davis retired from Brigham Young University in 1995, the project continued under the leadership of Professor Joseph W. Dellapenna of the Villanova University School of Law. In 1997, the Regulated Riparian Model Water Code was published. See Joseph W. Dellapenna (ed.), The Regulated Riparian Model Water Code, 1997. The Appropriative Rights Model Water Code is presently in the final stages of review leading to a committee publication.

Early in the project formulation and development process, the Committee recognized that effective water allocation and management necessitated planning and regulation on a water basin basis and acknowledged the need for a companion Model Code for the Shared Use of Transboundary Water Resources (SUTWR), or a model code for utilization of waters flowing across or along the boundary of sovereign governments.

Consequently, in 1995, ASCE initiated the SUTWR Project. The purpose of the project was to review transboundary agreements and develop a model agreement for utilization of water by sovereign governments or sub-units within sovereign nations. The model agreement was to focus on the allocation and use of shared waters and the resolution of conflicts involving such waters. The goal was to provide an agreement that would limit potential conflict while providing an appropriate balance between efficient use of the water resource for economic purposes, public health and ecological protection. The scope of the model was to include international agreements, interstate compacts and state-tribal agreements for regulatory purposes along or across political boundaries. It was to apply to any sharing of waters between independent political governments.

The Need for a Model Water Sharing Agreement

To appreciate the need for a model water sharing agreement, the Task Committee considered the problems arising from the use of shared water resources. When water resources are shared by two or more autonomous, independent governments, the timing and magnitude of the respective individual uses have been continual sources of conflict. The deficiency of adequate supplies of water is evident throughout much of the western United States, and the use of shared water resources is a major source of conflict. The interstate and international conflicts over the allocation of the waters of the Colorado River began early in the Twentieth Century and have still not been totally resolved. Even when water is relatively plentiful, the increasing demand for water from shared resources is present and has resulted in conflict. This has been graphically shown by the dispute between Florida, Alabama and Georgia over allocation of the waters of the Apalachicola-Chattahoochee-Flint River basin. The problem is pervasive. Few river basins in the continental United States are contained within a single state's boundaries.

The problem is magnified in the international arena. There are 268 major rivers shared between and among two or more nations. These international rivers cover almost one half of the total land surface of the globe. Fifty-three rivers are shared by three or more nations, with the Danube being shared by 17 riparian countries. International river basins sustain over 40% of the world's population. Almost 25% of the world's population live in the earth's semiarid and arid zones where scarcity of water is often acute. The potential for conflict is enormous. Among others, protracted conflict over shared waters exists between Turkey, Iraq and Syria in the Tigris-Euphrates basin; between Jordan and Israel regarding their opposite bank sharing of the Jordan River; and between nations in the Nile River Basin. The Ganges River is a source of dispute between India and Bangladesh. Armed conflict has occurred between Ecuador and Peru over the Cenepa River. The breakup of the Soviet Union has caused conflict between former members, especially in the arid regions east of the Caspian Sea. Seven active transboundary water disputes currently exist in Africa; 6 in Europe and Asia; and at least 5 in the Americas.

The need for effective co-operation among riparian countries is greater now than ever before because of the growing demand for water in various co-basin countries and the increasingly harmful effects of activities in upstream countries. While some form of interstate compact covers most of the shared river basins in the United States, many were drafted in the first half of the Twentieth Century and were uni-dimensional and limited in scope, being oriented to specific problems. Since water resource experts have recognized that the shared use of water resources is most effective when management is on the river basin level and when management of the shared resource is comprehensive and multi-dimensional, it can be argued that many interstate water compacts are inadequate to resolve the more complex water sharing issues presented at the dawn of the Twenty-First Century. Internationally, the problems are more acute. Over a third of the 200 international river basins are not covered by any international agreement, and only some 30 have truly co-operative institutional arrangements. The need exists for a model code that presents clear standards and principles for effective and efficient water sharing between autonomous political entities.

The ASCE SUTWR Project

The ASCE SUTWR Project sought to fill this void. Advice and assistance concerning the management of shared water resources were solicited from engineers engaged in water resources development, from government administrators working with water from a variety of perspectives, from lawyers representing development interests and from lawyers representing environmental interests, from business people representing a wide spectrum of industries, from academics in disciplines including civil engineering, economics, hydrology, law, and political science, and from environmental activists. A number of experts from such varied backgrounds gave detailed critiques of the several drafts of the project; many of these also attended two or more meetings per year where the drafts were discussed in detail. Those involved in the project agree that overall the end products are carefully balanced to represent a coherent body of law that would markedly improve the management of shared water resources.

Stephen E. Draper chaired the working group that drafted the Model Code for the Shared Use of Water Resources. Members of the working group included William E. Cox; Joseph W. Dellapenna,; J. Wayland Eheart; Christopher Estes; Conrad G. Keyes, Jr.; Kris G. Kauffman; Olen Paul Matthews; Zachary McCormick; Don Phelps; and Gerald Sehlke. Bob Chuck, Ron Cummings, N.D. "Skeeter" McClure and R. Peter Terrebonnne provided significant assistance. This document is the final report of the Model Agreements for the Shared Use of Transboundary Water Resources. The final draft of this report was subject to independent review by three prominent experts in interstate compact and the law of transboundary water sharing who had not been actively involved in the drafting of the Model Agreements. They were Professor Jim Bross of Georgia State University School of Law; Marshall Goulding, former Chief Engineer of the Susquehanna River Basin Commission; and Richard H. Cox, retired Commissioner of the Hawaii State Commission on Water Resources Management.

In the formulation of the model agreement, the SUTWR Committee identified the principal issues that limit effective transboundary water sharing. Although water scarcity and the increasing competition for water suggest that comprehensive management of a shared river basin is appropriate, a significant challenge to overcome is the prevailing tendency for governments at all levels to resist outside control over and interference with their internal affairs and those decisions which affect economic growth or quality of life. Each sovereign Party, whether it be a nation-state or a state within a federal system or tribal entity within a national federal system, wants to maintain authority over the people, places, assets and natural resources within its political boundaries. Local control is the operative word.

Strong incentives are required for a sovereign government to relinquish control to others. Early in the process the committee recognized that the degree to which the Parties were willing to relinquish sovereign control over water resources depended on the hydrologic, geographical and political situation. In some situations the Parties may relinquish considerable autonomy in the search for efficient water management. In other situations the Parties insist on maintaining control of the waters within their boundaries, being content to coordinate water management activities in some manner. Therefore, the committee chose to develop three distinct model transboundary agreements to serve as a framework for individual agreements.

One model agreement, Model A (Coordination and Cooperation), provides the framework for agreements in situations where each Party prefers to retain complete management authority over that portion of the shared water resource that lies within its boundaries. However, the Parties accepting Model A must acknowledge that effective management of those resources requires significant cooperation and coordination. They recognize a mechanism is needed for exchanging water-related information concering shared waters, to include hydrologic data and proposed projects that may affect extra territorial waters. Model A makes clear provision for the sovereign Parties to adequately protect their individual interests while coming to the table to resolving, where possible, transboundary water issues. The model can be described as the "State's Rights" model. This model insures that each Party is aware of the quantity and quality of water that will be available for its use. It provides for the Parties to share information about the activities and plans that may affect that availability. This "cooperation/coordination" model, Model A, makes clear provision for Sovereign Parties to protect their individual interests while resolving transboundary water issues peacefully.

A second model agreement, Model B (Limited Purpose), is concerned with water management needs that are limited in scope. It is a model wherein the Parties enter into an agreement to achieve certain limited purposes. The willingness to concede authority over management of internal water resources is limited to what is necessary to achieve a specific goal. The limited purpose goals may vary from simple allocation of water released from a single reservoir, to prioritization of needs during droughts, to other matters such as water quality control.

An interesting water management strategy may be developed using Model A as the framework for identifying individual water resources projects that may be defined by separate individual agreements which are based on Model B. The 1909 *Treaty between the United States and Great Britain relating to Boundary Waters* is an example of such a water management strategy. The Treaty itself provides the mechanism for coordination and cooperation between the two parties while specific side agreements have been developed, such as the 1978 *Great Lakes Water Quality Agreement*, the 1948 *Agreement to establish the International Souris-Red River Engineering Board* to allocate the waters of those rivers and the 1941 *Agreement on the International Columbia River Board* to monitor the effects of the operation of Grand Coulee Dam and Roosevelt Lake on the levels and flows of the Columbia at the international boundary.

The third model agreement, Model C (Comprehensive Management), provides a model for comprehensive planning and management of shared water resources. It is the model that best conforms to efficient water management according to science-based management and is the model most of the experts on the working group recommend. In this model, the Parties share authority over the water resource at issue. This model is appropriate for those situations in which the sovereign Parties wish to achieve optimal use of the shared water resources by multidimensional management on a river basin basis. The agreement is extensive and considers all aspects of management of the water resources. The goal of this comprehensive, integrated agreement is to achieve allocation that is equitable and fair to all prospective users. It requires that each Party restrict practices to the reasonable use of water and provide sufficient data to the other Parties to verify that its use is reasonable. Prime examples of Model A are the 1961 *Delaware River Basin Compact* and the 1970 *Susquehanna River Basin Compact*.

Regardless of the model chosen, however, a final word of caution is extended to the Parties drafting the agreement. On many occassions each source of water, whether it be surface, underground or atmospheric water, is assessed and managed as separate and distinct source from the others. However, such individualized assessments do not provide an integrated understanding of either the water sources or their use. Prior to implementing a transboundary water use apportionment or allocation, participants should conduct a comprehensive water resources assessment. The assessment should evaluate:

- The sources of water;
- The volume, flow, and distribution of water within the basin;
- Water quality, and
- Water demand, including environmental and anthropogenic need.

The goal should be to determine the hydrologic mass balance for the basin, and to establish a detailed conceptual model of the basin which defines the water resources available for use, carefully defining the time period and area covered and present and future water demands.

A comprehensive assessment requires that three conditions be met. First, the Parties must define the critical hydrologic conditions that frame the water sharing agreement. Second, adequate and reliable data concerning the water sources and demands must be available. Third, a shared water resource database should be developed and maintained. Certain parameters define the framework of the agreement. The Parties should specify the range of hydrologic events that the agreement will cover. This means the Parties must clearly establish the quantitative measures of "normal" hydrologic conditions and establish the levels when special management for drought (dry) or (wet) conditions arise. The Parties must also define the levels of water quality degradation they are willing to accept as a result of meeting the demands for the water.

Adequate and reliable data is required. What constitutes adequate and reliable data depends on the water sources and needs of the specific individual basins and the willingness of the Parties to expend the necessary resources to collect, catalog, and make available the data. The data should be sufficient to determine the water available under drought and flood conditions as well as normal conditions. A understanding of the interaction between surface water and ground water is essential.

A number of factors influence the availability of basin water. These should be identified and examined through a review and analysis of climatology, physiography, geology and existing underground and surface water resources, including reservoirs, and interaction of these factors. The search for water to satisfy needs and demands involves quantification of potential sources and losses.

Form And Sources

These model agreements follow the form commonly used today in the drafting of proposed uniform state laws under the auspices of the National Conference of Commissioners of Uniform Laws. That form consists of a statutory language in bold face that a legislature could enact with or without change. This language is arranged in sections, generally consisting of a single sentence, for ease of citation. The numbering of the sections consists of three parts, indicating the chapter of the Code, the part of the chapter, and the sequential numbering of each section within that part. For example, §2A-3-01 means Section 1, Part 3, Article 2 of Model A, the Model Agreement for Coordination and Cooperation in the Management of Shared Water Resources.

Each Section is arranged into four segments. The first segment presented in bold lettering, provides the statutory language of the Section itself. This is followed by a Commentary segment that amplifies the statutory language and provides the committee's reasoning for including the statutory language. This Commentary is followed by a segment that provides cross-references. Finally, a fourth segment provides a list of existing agreements that contain similar provisions.

Each section necessarily is discretionary in that negotiators drafting the agreement, even were they to decide to enact the bulk of a particular Model Agreement, could delete or change any particular section. Nonetheless, the drafters of this Code attempted to create a complete, comprehensive, and well integrated statutory scheme for creating or refining a Model Agreement for the Shared Use of Transboundary Water Resources capable of dealing with the water management problems of the twenty-first century. The drafters have concluded that nearly every section of this Code is necessary to achieve that goal. Where the committee determined that the purpose for the particular Section could be achieved by alternative statutory language, alternatives are presented. In a limited number of Sections, the committee determined that the suggested statutory language was not appropriate for all transboundary water sharing agreements. For instance, §1-1-05 (national security) is appropriate only for those agreements of an international character, not for interstate compact purposes. Consequently it is specifically noted as "Optional, for international use." Several other sections (e.g., §4A-3-5, §4A-3-6, §4A-3-7, §4B-1-05, §4B-1-06, §4B-1-07, §4C-3-5, §4C-3-6, §4C-3-7) are also specifically noted "optional."

These model agreements refer to current *ASCE Policy Statements*, as of the date of this publication, and to certain common references. For the basic law regarding interstate water allocation, the central source is Douglas L. Grant, *Interstate Water Allocation* ["Grant"] in 4 Waters and Water Rights, chs. 43-48 (7 vols., Robert E. Beck ed., 1991 w. 1997 Cumulative Supplement. For the basic international law of the non-navigational use of shared water resources prior to enactment of *the Law of the Non-Navigational Uses of International Watercourses*, the central source is Albert E. Utton, "International Waters" ["Utton"] in 5 Waters and Water Rights, chs. 49-51. For the Colorado River Basin Treaty and compacts, the central source is Lawrence J. MacDonnell, "Colorado River Basin" ["MacDonnell"] in 6 Waters and Water Rights, pp. 5-55. For the Delaware and Susquehanna River Basins River Basins and Compacts, the central source is Joseph W. Dellapenna, "Delaware and Susquehanna

River Basins" ["Dellapenna 1"], in 6 Waters And Water Rights, pp. 137-147. The primary source for active transboundary water disputes is Paul Samson and Bertrand Charrier in International Freshwater Conflict: Issues and Prevention Strategies, 1997.

Legal Evolution of Shared Water Agreements

To use any of the model agreements effectively, the context and scope of the legal regime in which the agreement will operate must be understood. As a first step, each of the agreements was formulated and developed for applicability to the legal regime available in the United States for resolving interstate or tribal-state water disputes. Review of the existing interstate water allocation compacts in the United States revealed the strengths and weaknesses of the scope of each agreement and the effectiveness of individual provisions. This review developed the foundations for each of the models.

As a second step an extensive review of international agreements was undertaken to verify the efficacy of some provisions and, in other cases, improve on the provisions developed from U.S. experience. The result is three water sharing agreements that can be used directly within the U.S. context but which may be used with minor alteration in the international context. However, the Committee notes that, with the "globalization" of economic and social forces, it can be expected that the form and substance of interstate compacts and international water sharing agreements will converge.

Water sharing agreements between or among sovereign states must conform to the legal structure of the river basin for which which they are drafted. Consequently, interstate compacts within the United States need not contain some of the procedural safeguards that are necessary for international water sharing agreements. A review of both American and international law is appropriate for an understanding of the applicability of the model codes and the extent their modification may be necessary.

Within the United States, the Federal-State partnership established by the U.S. Constitution does provide a stable legal relationship between the states upon which interstate conflict resolution may be based. Three categories of sovereignty arise under the constitutional powers of the federal, state and tribal governments. Under federalism, the states as sovereign entities maintain control over inland waters, with some important exceptions. These exceptions include specific legislation, such as the Clean Water Act, 33 U.S.C. §§ 1251 *et seq* (CWA)and other legislation, and certain legal doctrines established by the Courts. Such legislation and doctrines include federal reserved water rights that attach to federal lands in the west, federal control over navigability, and federal supremacy for federal waterpower projects. Independent of the states, Indian tribes also may assert sovereignty over certain waters as the result of the treaties the various tribes entered into with the United States. Thus, under some conditions, the state controls use of waters within its boundaries. Under other conditions, an Indian tribe may control certain waters within a state. Finally, under still other conditions, the Federal government may control certain waters within a state. Added to this "brew," the use of the waters within one state may significantly affect the manner or quantity of waters within another state. Inevitably, conflicts over use of the water develop. Significantly, the Constitution provides a mechanism for conflict resolution, and a forum for final conflict resolution exists in the U.S. Supreme Court.

There are two major methods of resolving interstate water allocation disputes: (1) U.S. Supreme Court "original jurisdiction" suits between the states and (2) interstate water apportionment compacts. The three Model SUTWR agreements presented in this document relate primarily to the latter method, interstate water apportionment compacts. However, an understanding of the other method, U.S. Supreme Court "original jurisdiction" suits between the states, is necessary since any interstate compact must arguably conform to the principles of interstate water sharing as established by the Supreme Court. Additionally, it should be recognized that compact interpretation would be accomplished by the federal courts according to those principles laid out by the Supreme Court in its federal interstate common law doctrine of "equitable apportionment."

The basis of the "equitable apportionment" doctrine is that interstate waters should be apportioned by balancing equities. Equity (or fairness) is achieved by balancing the stake (or value) each Party may have in utilizing the shared resource. The factors to be analyzed to determine a "fair" allocation of water between or among the Parties include:

(1) Physical and climatic conditions;
(2) Consumptive use of water;
(3) Character and rate of return flows;
(4) Extent of established uses and economies built on them;
(5) Availability of storage water;
(6) Practical effect of wasteful uses on downstream area;
(7) Damage to upstream areas compared to the benefits to downstream areas if limitations were to be imposed on the former.

Theoretically, an interstate compact can apportion waters in any manner or procedure the party states choose. However, the Supreme Court has stated that, to the extent an interstate compact is based on equitable apportionment principles, the compact will preempt contrary state laws. This statement arguably leads to the conclusion that any interstate water allocation compact should meet the admittedly vague "equitable apportionment" standards.

Within the international arena, the problems with effective conflict resolution are magnified for two reasons. First, no forum for final conflict resolution of water disputes exists in relation to water sharing conflicts. The United Nations, the World Court and the World Trade Organization (WTO), provide forums for conflict resolution in some subject matter and under certain conditions, but no forum for conclusive judgment for conflicts involving water sharing exists. Secondly, no clear standards exist in customary international law that relate to transboundary water sharing.

Under long-standing customary international law, a sovereign state had the exclusive rights to all natural resources within its boundaries. This customary rule of international law recognized undisputed ownership of water within political boundaries. Necessarily, this rule presents the potential for serious conflict over shared water resources, especially in upstream-downstream disputes.

Over the years, various international conferences and congresses have recognized that this absolute control philosophy was not an acceptable means of resolving conflicts. However, no definitive international treaty or agreement has been ratified on a worldwide basis to change this rule. With no such global agreement in place, a legal question emerges whether this precept of customary international law remains valid or whether a series of recent international conferences and agreements have changed the law.

The first serious challenge to this customary rule of law appeared in 1966. The 1996 *Helsinki Rules on the Use of the Waters of International Rivers* presented a consensus of the attending States that each co-riparian is entitled to "a reasonable and equitable share in the beneficial use of the waters of an international drainage basin." Since 1966, a variety of other sources of international law have presented conflicting views. The 1972 *Stockholm Declaration on the Human Environment* articulated the view that sovereign states have "the responsibility to ensure that activities within their jurisdiction or control do not cause damage to the environment of other States or of areas beyond the limits of national jurisdiction." Standing by itself, this rule could clearly supersede the customary rule of law that recognizes undisputed ownership of water within political boundaries. However, a companion declaration in this 1972 Declaration also articulated the need to protect "the sovereign right [of States] to exploit their own resources pursuant to their own environmental policies."

The 1978 *Amazon Pact Treaty* proclaims "that the exclusive use and utilization of natural resources inside their own territory is a right inherent to the states' sovereignty and that its exercise will not be subject to restrictions other than those imposed by international law …" The standards of international law were not described however. The 1992 *United Nations' Convention on the Protection and Use of Transboundary Watercourses and International Lakes* declared that the parties "shall take all appropriate measures to prevent, control and reduce any transboundary impact ... (and) ensure that transboundary waters are used in a reasonable and equitable way, taking into particular account their transboundary character, in the case of activities which cause or are likely to cause transboundary impact..." Oriented to European transboundary water quality, this Convention calls for "sustainable water-resources management" but does not discuss allocation of water.

In 1992 the United Nations Commission on Sustainable Development reiterated the sovereign right to exploitation of internal resources. *See* Report of the United Nations Conference on Environment and Development (Rio de Janeiro, 3-14 June 1992). This report, published as Agenda 21, stated to the Secretary-General that

> States have, in accordance with the Charter of the United Nations and the principles of international law, the sovereign right to exploit their own resources pursuant to their own environmental policies and have the responsibility to ensure that activities within their jurisdiction or control do not cause damage to the environment of other States or areas beyond the limits of national jurisdiction.

Clearly, without a definitive international treaty, signed and ratified by a significant number of the world's powers, current customary law concerning the non-navigational use of international watercourses is still ambiguous. Arguably, the standards for the non-navigational use of international watercourses has changed little from its position set by the customary rule of international law that recognizes undisputed ownership of water within political boundaries.

The 1998 U.N. *Convention on the Law of the Non-Navigational Uses of International Watercourses* may end this confusion about customary law. The U.N. General Assembly approved this Convention in May 1998 by a vote of 104-3. Although it will not enter into force until ratified by 35 nations, the Convention may provide clear standards for agreements concerning the shared use of international watercourses. It will institute the standard of "equitable and reasonable utilization" for use of transboundary waters. The Convention states:

Article 5

Equitable and reasonable and participation

1. Watercourse States shall in their respective territories utilize an international watercourse in an equitable and reasonable manner. In particular, an international watercourse shall be used and developed by watercourse states with a view to attaining optimal and sustainable utilization thereof and benefits therefrom, taking into account the interests of the watercourse States concerned, consistent with adequate protection of the watercourse.

2. Watercourse States shall participate in the use, development and protection of an international watercourse in an equitable and reasonable manner. Such participation includes both the right to utilize the watercourse and the duty to cooperate in the protection and development thereof, as provided in the present convention.

The Convention's doctrine of "equitable and reasonable utilization" is similar to, but more expansive than the U.S. Supreme Court 's doctrine of "equitable apportionment."

Article 6

Factors relevant to equitable and reasonable utilization

1. Utilization of an international watercourse in an equitable and reasonable manner within the meaning of article 5 requires taking into account all relevant factors and circumstances, including:

   (a) Geographic, hydrographic, hydrological, climatic, ecological and other factors of a natural character:

(b) The social and economic needs of the watercourse States concerned;

(c) The Population dependent on the watercourse in each watercourse State;

(d) The effects of the use or uses of the watercourses in one watercourse State on other watercourse States;

(e) Existing and potential uses of the watercourse;

(f) Conservation, protection, development and economy of use of the water resources of the watercourse and the costs of measures taken to that effect;

(g) The availability of alternatives, of comparable value, to a particular planned or existing use.

2. In the application of article 5 or paragraph 1 of this article, watercourse States concerned shall, when the need arises, enter into consultations in a spirit of cooperation.

3. The weight to be given to each factor is to be determined by its importance in comparison with that of other relevant factors. In determining what is a reasonable and equitable use, all relevant factors are to be considered together and a conclusion reached on the basis of the whole.

Other significant provisions of the Convention include the requirement for all Parties to "prevent the causing of significant harm to other watercourse States" (Article 7). It declares that all watercourse Parties must "cooperate on the basis of sovereign equality, territorial integrity, mutual benefit and good faith in order to attain optimal utilization and adequate protection of an international watercourse" (Article 8).

Notably, the U.N. *Convention on the Law of the Non-Navigational Uses of International Watercourses* applies only to co-riparians on international watercourses. Attention is directed to the terminology used in the *Law of the Non-Navigational Uses of International Watercourses*, especially regarding the term "equitable and reasonable utilization." Nowhere in the Convention does the term "beneficial use" appear. This term of art is used extensively in those U.S. states adhering, in whole or in part, to the Prior Appropriation water allocation doctrine. Because the term "beneficial use" raises significant policy issues, especially with regard to the classification of instream uses as valid and legally protected water uses, and because current international law uses the term "equitable and reasonable utilization" to denote those water uses that are permissive, the latter phrase is used in this Model Agreement.

The significant difference between international law and the laws of the United States is that, for the United States, the U.S. Supreme Court provides the institutional legal framework with a "court of last resort" to resolve interstate water disputes. No such forum exists for international water disputes; reliance must be placed on internationally accepted legal norms such as a ratified *Convention on the Law of Non-Navigational Uses of International Watercourses*.

Conclusion

The purposes of the American Society of Civil Engineers SUTWR project were to review and analyze existing transboundary agreements, both within the U.S. Federal system and internationally, and develop a model transboundary water sharing agreement for use by sovereign governments. Because of the nature of sovereignty and its attendant limitations for full efficiency in water use, three model agreements were developed for use according to the degree of willingness of the Parties to forego sovereignty. The models are intended for use in international agreements, interstate compacts or state-tribal agreements for regulatory purposes along or across political boundaries. All three models focus on the allocation and use of shared waters and on resolving conflicts involving such waters. They seek to limit potential conflict while providing an appropriate balance between efficient use of the water resource for economic purposes, public health and ecological subsistence.

Contents

Model A..1
    Model Agreements for Coordination and Cooperation in the Management of
    Shared Transboundary Water Resources

Model B..38
    Limited Purpose Agreement Concerning the Shared Use of Transboundary
    Water Resources

Model C..79
    Comprehensive Management Agreement Relating to the Shared Use of
    Transboundary Water Resources

Index 181

MODEL AGREEMENT FOR THE COORDINATION AND COOPERATION IN THE MANAGEMENT OF SHARED TRANSBOUNDARY WATER RESOURCES

Model A

Contents

Model Agreement for the Coordination and Cooperation in the Management of Shared Transboundary Water Resources

Introduction 4
Article 1A – Declaration of Policies and Purposes 6
General Policies 6
Coordination and Cooperation 8
Good Faith Implementation 9
Preservation of Federal Rights (Optional, for U.S. Use) 10
National Security (Optional, for International Use) 11

Article 2A – General Provisions 11
Part 1: General Obligations 11
Effective Date 11
Consent to Jurisdiction (Optional, for U.S. Use) 13
Duration to Agreement (Optional) 13
Powers of Sovereign Parties; Withdrawal (Optional, International Use) 14
Amendments and Supplements (Optional) 15
Existing Agencies (Optional) 15
Severability 16
Annexes 16
Part 2: Definitions 17
_____ Basin 17
Drought 17
Flood 17
Interbasin Transfer 18
Party or Parties 18
Waters of the Basin 18

Article 3A – Administration 19
Part 1: Administrative Authority 19
Alternative 1 (Administration by Party Officials) 19
Use of Water Management Officials of the Parties 19
Substitution of Officials 20
Implementation and Verification of Agreement 20
Funding 21
Alternative 2 (Administration by Commission) 21
Commission Created 21
Commission Authorities and Responsibilities 23
Commissioners 23
Status, Immunities and Privileges (Optional) 25
Commission Organization and Staffing 26
Advisory Boards (Optional) 27
Rules of Procedure 28
Part 2: Powers and Duties 28
General Powers and Duties 28
Special Powers and Duties (Optional) 30
Part 3: Administrative Procedures 31
Meetings, Hearings and Records (Optional) 31
Funding and Expenses of the Commission 32

Article 4A – Coordination of Water Issues 33
Exclusive Jurisdiction and Control 33
Data Exchange 34

Article 5A – Dispute Resolution (Optional) 36
Resolution by Signatory Parties 36
Right to Litigate 37

Signatures 37

INTRODUCTION

Model A. the Agreement for Coordination and Cooperation in the Management of Shared Water Resources, provides a reference to be used by States to facilitate the exchange of data and other information pertinent to independent water planning and development by the respective Parties. In many situations involving the sharing of water between, along, and across political borders, the Parties are not prepared to relinquish their sovereign rights and duties over the waters within their boundaries but acknowledge that effective management of those resources cannot be accomplished without significant cooperation and coordination. This model transboundary agreement is designed to provide a mechanism for coordination and cooperative that insures each Party is aware of the existing quantity and quality of water available for use as well as the activities and plans of the other Party that may impact on that availability. It provides a means for exchanging boundary water-related information, including hydrologic data and information on proposed projects that may affect extraterritorial waters. This "cooperation/coordination" model makes clear provision for Sovereign Parties to adequately protect their individual interests while resolving, as effectively as possible, transboundary water issues and conflicts.

This Agreement is partly based on the *Treaty between the United States and Great Britain relating to Boundary Waters*, 36 Stat. 2451 (1909). This Treaty manages transboundary water sharing issues across and along the longest undefended boundary between two countries in the world (5,400 miles). It has been used over the past nine decades to successfully resolve over 95% of the issues addressed under its jurisdiction. The Treaty defines the transboundary waters and the existing rights within each Party's jurisdiction as well as certain preexisting or planned conditions under which the Treaty is effective. It sets forth a process and an administering body for future decision-making with regard to issues of the defined transboundary water resources. The International Joint Commission (IJC) established by the treaty sets forth Rules of Procedure for its operation as provided by the treaty. Certain provisions of this Treaty are, however, inappropriate for inclusion in water sharing agreements for other water basins. Modifications and revisions to the Treaty have therefore been made to a number of provisions before inclusion in the Model Agreement. These modifications allow the Model A to be sufficiently flexible for use on both an interstate and international scale and in a variety of geopolitical settings. These modifications and revisions resulted from an evaluation of a number of successful water sharing agreements as well as manuscripts, treatises and other studies published by experts and learned scholars.

International agreements provided significant assistance in developing policies and provisions that are science-based and which reflect the globalization of water policy planning. From a policy standpoint, this Model Agreement for Coordination and Cooperation in the Management of Shared Water Resources conforms to the United Nations *Convention on the Law of the Non-Navigational Uses of International Watercourses*, United Nations Document A/51/869 (1998), signed by the United States as one of the 104 signatories. It has not yet been ratified.

The Convention recommends the establishment of the dual criteria of "equitable and reasonable utilization" of the water resources and the need to "exchange data and consult on the possible effects of planned measures on the condition" of the water resource. This Model Agreement, as well as other international agreements, is relevant to interstate agreements because of what it offers as guiding principles for the peaceful resolution of water disputes. Much international study has been concentrated on developing successful water sharing agreements and the states within the United States can profit from this study.

AGREEMENT CONCERNING THE WATERS SHARED BETWEEN

_____ AND _____

ARTICLE 1A

DECLARATION OF POLICIES AND PURPOSES

§1A-1-01 GENERAL POLICIES

(a) The water resources of the _____ River Basin have local, regional and national significance and their management, development, and control by the individual Parties and under appropriate arrangements for intergovernmental cooperation are public purposes for the respective signatory Parties.

(b) The major purposes of this Agreement are to provide the coordination and cooperation necessary for the management and development of _____ River and its tributaries, the exchange of data and other information pertinent to independent water planning and development by the respective Parties of the _____ River and its tributaries, to promote interstate comity, and to remove causes of present and future controversy.

(c) The general policies of this Agreement include facilitating the equitable and reasonable use of the water resources shared between the Parties, the exchange of data and other information pertinent to water utilization by the respective Parties, and the cooperation and consultation necessary to achieve the purposes of this Agreement.

(d) The bases for this Agreement are the geophysical, climatic, meteorological and other conditions peculiar to the _____ River Basin, and application of its provisions is limited to those waters.

Commentary: Economic growth and prosperity require adequate supplies of quality water on a regular and sustained basis. This requires that utilization of shared waters be well coordinated among or between the Parties sharing the waters. An effective Agreement can facilitate adequate planning, conservation, utilization, development, management and control of water resources on a water basin basis, in a manner that is reasonable and equitable under the circumstances and that causes no significant harm to other Parties. A key challenge for the parties is to make more efficient and productive use of water and to reshape the water policies of the individual Parties to better respond to periods of water shortages. (*See* S. Postel, Dividing the Waters: Food Security, Ecosystem Health, and the New Politics of Scarcity, (1996)).

One criterion of the *Convention on the Law of the Non-Navigational Uses of International Watercourses*, United Nations Document A/51/869 (1998), makes the definition of the waters to which an agreement applies a mandatory provision in the agreement. The Parties must carefully frame the extent of the specific water resources involved in the agreement. They should identify the type and geographical extent of the water resources to be subject to the Agreement. In order to formulate an effective agreement, the Parties should analyze the factors that influence the water resource in question, including the climatology, physiology, geology and the interaction between underground and surface water resources. The analysis must identify pollution sources and the resulting impact on water quality. The geographic scope of the water resources to be covered by the Agreement should be sufficiently expansive to fully address all the water sharing issues involved. The use of the term "_____ River Basin" if objectionable for any reason by one or more Parties, may be changed to "_____ River and its tributaries," "boundary waters," "shared waters," "frontier waters" or any phrase that accurately describes and encompasses the entirety of the water resources subject to the Agreement

It is important to acknowledge that the Agreement reflects the particular circumstances and compromises reached in its formulation, and that it applies only to the waters shared between the Parties. The inclusion of §1A-1-01(b) may avoid later claims that other rivers and streams, or other bodies of water, are subject to the agreement.

Cross-references: §1A-1-02 (coordination and cooperation); §1A-1-03 (good faith implementation); §1A-1-04 (preservation of federal rights); §2A-1-01 (effective date); §2A-1-02 (consent to jurisdiction);§2A-1-03 (duration of agreement); §2A-1-04 (amendments and supplements); §2A-1-05 (powers of sovereign parties; withdrawal); §2A-1-06 (existing agencies); §2A-1-07 (limited applicability); §2A-1-08 (annexes); §2A-2-01 (____ Basin); §2A-2-05 (party or parties); §3A-1-01 (alternative 1, use of party officials); §3A-1-02 (alternative 1, substitution of officials); §3A-1-03 (altrnative 1, implementation and verification of agreement); §3A-1-04 (alternative 1, funding); §3A-1-01 (alternative 2, commission created); §3A-1-02 (alternative 2, commission jurisdiction); §3A-1-03 (alternative 2, commissioners); §3A-1-04 (alternative 2, status, immunities and privileges); §3A-1-05 (alternative 2, commission organization and staffing); §3A-1-06 (alternative 2, advisory boards); §3A-1-07 (alternative 2, rules of procedures); §3A-2-01 (alternative 2, general powers and duties); §3A-2-02 (alternative 2, special powers and duties); §3A-3-01 (alternative 2, meetings, hearings and records); §3A-3-02 (alternative 2, funding and expenses of the commission); §4A-1-01 (exclusive jurisdiction and control); §4A-1-02 (data exchange); Article 5A (dispute resolution).

Similar Agreements: *Pecos River Compact*, 63 Stat. 159 (1948); *Delaware River Basin Compact*, Pub. L. 87-328, 75 Stat. 688 (1961); *Susquehanna River Basin Compact*, Pub. L. No. 91-575, 84 Stat. 1509 (1970); *Alabama-Coosa-Tallapoosa River Basin Compact*, O.C.G.A. 12-10-110 (1997); *The Apalachicola-Chattahoochee-Flint River Basin Compact*, O.C.G.A. 12-10-110 (1997); *Agreement Between the People's Republic of Bulgaria and the Republic of Turkey Concerning Co-operation in the Use of the Waters of Rivers Flowing through the Territory of Both Countries*, UNTS, Vol. 807, 117 (1968); *Convention on the Law of the Non-Navigational Uses of International Watercourses*, United Nations Document A/51/869 (1998).

§1A-1-02 COORDINATION AND COOPERATION

The parties agree

(a) To cooperate and consult with the other Parties to this Agreement in their development and utilization of the water and related resources of the waters shared by the Parties in order to ensure use of those waters while minimizing harm to other Parties.

(b) To cooperate on the basis of sovereign equality and territorial integrity in the utilization and protection of the shared water resources.

(c) To conduct themselves with an absence of malice and deceit, and with no intention to seek unconscionable advantage.

Commentary: This provision provides a framework for the Parties in their development of their individual water policy planning. It recognizes that each Party must follow in its rationale management of water resources certain fundamental principles. These general objectives and principles improve the likelihood of accomplishing the purposes of the agreement.

A nation normally enters into any international agreement with a position of self-interest. In the negotiations, each Party seeks the rights and authority critical to certain political, economic or social objectives while ceding less critical rights and authority to the other nations. While accepting this fact, all Parties have a duty to cooperate and negotiate in good faith. This principle is the foundation of international law, and it applies in all relations between sovereign states.

Cross-references: §1A-1-01 (general policies); §2A-1-06 (existing agencies); §3A-1-01 (alternative 1, use of party officials); §3A-1-02 (alternative 1, substitution of officials); §3A-1-03 (alternative 1, implementation and verification of agreement); §3A-1-04 (alternative 1, funding); §3A-1-01 (alternative 2, commission created); §3A-1-02 (alternative 2, commission jurisdiction); §3A-1-03 (alternative 2, commissioners); §3A-1-04 (alternative 2, status, immunities and privileges); §3A-1-05 (alternative 2, commission organization and staffing); §3A-1-06 (alternative 2, advisory boards); §3A-1-07 (alternative 2, rules of procedures); §3A-2-01 (alternative 2, general powers and duties); §3A-2-02 (alternative 2, special powers and duties); §3A-3-01 (alternative 2, meetings, hearings and records); §3A-3-02 (alternative 2, funding and expenses of the commission); §4A-1-01 (exclusive jurisdiction and control); §4A-1-02 (data exchange); Article 5A (dispute resolution).

Similar Agreements: *Agreement Between the People's Republic of Bulgaria and the Republic of Turkey Concerning Co-operation in the Use of the Waters of Rivers Flowing through the Territory of Both Countries,* UNTS, Vol. 807, 117 (1968); *Convention Between Switzerland and Italy Concerning the Protection of Italo-Swiss Waters Against Pollution,* UNTS, Vol. 957, 277 (1972); *Stockholm Declaration of the United Nations Conference on the Human Environment,* 11 I.L.M. 1416 (1972); *Treaty for Amazonian Cooperation,* 17 ILM 1046 (1978); *Convention Between the Federal Republic of Germany and the Czech and Slovak Federal Republic and the European Economic Community on the International Commission for the Protection of the Elbe,* International Environmental Law, Multilateral Agreements, 976:90/1 (1990); *Convention on the Protection and Use of Transboundary Watercourses and International Lakes,* 31 I.L.M. 1312 (1992); *The North American Agreement on Environmental Cooperation between the Government of the United States of America, the Government of Canada, and the Government of the United Mexican States,* 32 I.L.M. 1480 (1993); *Treaty of Peace between the State of Israel and the Hashemite Kingdom of Jordan,* 34 I.L.M. 43 (1994); *Agreement on Cooperation for the Sustainable Development of the Mekong River Basin,* 34 ILM 864 (1995); *Convention on the Law of the Non-Navigational Uses of International Watercourses,* United Nations Document A/51/869 (1998).

§1A-1-03 GOOD FAITH IMPLEMENTATION

The Parties agree to implement immediately all provisions of this Agreement, and each Party covenants that its officers and agencies will not hinder, impair, or prevent any other Party from carrying out any provision of this Agreement.

Commentary. This provision complements the provision concerning the duty to cooperate and negotiate in good faith. It must be noted, however, that good-faith misinterpretation of compact obligations do not excuse a Party from damage liability. *See Texas v. New Mexico,* 482 U.S. 124 (1987). In that case, the Supreme Court reasoned that a compact is a contract, and standard contract law does not allow a defense based on misinterpretation of contract obligations. *See* Grant, §45.07(c), §46.05(d).

Cross-references: §1A-1-01 (general policies); §1A-1-02 (coordination and cooperation); §2A-1-06 (existing agencies); §2A-2-05 (party or parties); §3A-1-01 (alternative 1, use of party officials); §3A-1-02 (alternative 1, substitution of officials); §3A-1-03 (alternative 1, implementation and verification of agreement); §3A-1-04 (alternative 1, funding); §3A-1-01 (alternative 2, commission created); §3A-1-02 (alternative 2, commission jurisdiction); §3A-1-03 (alternative 2, commissioners); §3A-1-04 (alternative 2, status, immunities and privileges); §3A-1-05 (alternative 2, commission organization and staffing); §3A-1-06 (alternative 2, advisory boards); §3A-1-07 (alternative 2, rules of procedures); §3A-2-01 (alternative 2, general powers and duties); §3A-2-02 (alternative 2, special powers and duties); §3A-3-01 (alternative 2, meetings, hearings and records); §3A-3-02 (alternative 2, funding and expenses of the commission); §4A-1-01 (exclusive jurisdiction and control); §4A-1-02 ((data exchange); Article 5A (dispute resolution).

Similar Agreements: *Helsinki Rules on the Uses of the Waters of International Rivers,* 52 I.L.A. 484 (1966); *Stockholm Declaration of the United Nations Conference on the Human Environment,* 11 I.L.M. 1416 (1972); *Convention on the Law of the Non-Navigational Uses of International Watercourses,* United Nations Document A/51/869 (1998).

§ 1A-1-04 PRESERVATION OF FEDERAL RIGHTS (Optional, for U.S. use)

Nothing in this agreement shall be deemed:

(a) To impair or affect any rights or powers of the United States, its agencies or instrumentalities, in and to the use of the waters of the _____ River Basin nor its capacity to acquire rights in and to the use of said waters;

(b) To subject any property of the United States, its agencies, or instrumentalities to taxation by any Party, nor to create an obligation on the part of the United States, its agencies, or instrumentalities, by reason of the acquisition, construction or operation of any property or works of whatsoever kind, to make any payments to any State or political subdivision thereof, State agency, municipality, or entity whatsoever in reimbursement for the loss of taxes;

(c) To subject any property of the United States, its agencies, or instrumentalities to the laws of any State to an extent other than the extent to which these laws would apply in the absence of this *A*greement.

Commentary: These sections may be included in agreements between states of the United States. They are probably unnecessary to preserve federal rights, but inasmuch as Congress must approve the agreement, the inclusion of these provisions may make it easier to obtain that approval.

Cross References: §1A-1-01 (general policies); §2A-1-05 (powers of sovereign parties; withdrawal); §2A-1-06 (existing agencies); §2A-2-01 (_____ Basin); §3A-1-02 (commission jurisdiction).

Similar Agreements: *Rio Grande Compact of 1938,* 53 Stat. 785 (1938); *Republican River Compact,* 57 Stat. 86 (1943); *Belle Fourche River Compact,* 58 Stat. 94 (1944); *Pecos River Compact,* 63 Stat. 159 (1948); *Snake River Compact,* 64 Stat. 29 (1949); *Upper Colorado River Basin Compact,* 63 Stat. 31 (1949); *Yellowstone River Compact,* 65 Stat. 663 (1950); *Canadian River Compact,* 66 Stat. 74, (1952); *Klamath River Basin Compact,* 71 Stat. 497 (1957); *Bear River Compact,* 72 Stat. 38 (1955), amended 94 Stat. 4, Art. XIII (2) (1980).

§1A-1-05 NATIONAL SECURITY (Optional, for international use)

(a) Nothing in this Agreement shall be construed to require any Party to make available or provide access to information the disclosure of which it determines to be contrary to its essential security interests.

(b) Nothing in this Agreement shall be construed to prevent any Party from taking any actions that it consider necessary for the protection of its essential security interests relating to a formal declaration of war.

Commentary: National security concerns will necessarily take precedence over any program of water management and the exchange of data between the Parties. The exchange of data related to each Party's essential security interests would not be provided to the other Parties.

Cross-references: §1A-1-01 (general policies); §2A-1-05 (powers of sovereign parties; withdrawal); §2A-1-06 (existing agencies); §3A-1-02 (alternative 2, commission jurisdiction).

Similar Agreements: *The North American Agreement on Environmental Cooperation between the Government of the United States of America, the Government of Canada, and the Government of the United Mexican States*, 32 I.L.M. 1480 (1993).

ARTICLE 2A

GENERAL PROVISIONS

Part 1 General Obligations

§2A-1-01 EFFECTIVE DATE

Alternative 1: (For international use)

This agreement shall become operative when approved by the appropriate governing authorities of all Parties. The agreement will go into full force and effect at 12:01 a.m. [time zone] on the day immediately following the final act necessary for approval of the agreement, as defined by the domestic law of each Party.

Alternative 2: (For U.S. use only)

This agreement shall become operative when, subsequent to approval by the Legislature of each of the States, it is approved by the Congress of the United States by legislation providing, among other things, that:

    (a) Any equitable and reasonable uses hereafter made by the United States, or those acting by or under its authority, within a State, of the waters allocated by this agreement, shall be within the allocations hereinabove made for use in that State and shall be taken into account in determining the extent of use within that State.

    (b) The United States shall recognize, to the extent consistent with the best utilization of the waters for multiple purposes, that equitable and reasonable use of the waters within the Basin is of paramount importance to development of the Basin. This shall pertain to the exercise of rights or powers arising from whatever jurisdiction the United States has in, over and to the waters of the _____ River and all its tributaries. The United States government shall exercise no power that may interfere with the full equitable and reasonable use of the waters unless sucthe exercise of such power is in the interest of the best utilization of such waters for multiple purposes.

Commentary: Any agreement of this nature should specify the date or conditions upon which it will take effect. In the case of agreements between states of the United States, the conditions with respect to Congress are designed to provide some measure of protection against subsequent federal action that might disturb the allocation system agreed upon by the contracting Parties. Despite the requirement of federal approval of interstate compacts, the federal government is not normally a Party to those agreements and may not be bound by the provisions of those agreements unless there is specific legislation committing the federal government to be so bound. The provisions of Section 2-1-01, modeled after the *Republican River Compact*, 57 Stat. 86 (1943) and the *Bell Fourche Compact*, 58 Stat. 94 (1944) condition the effectiveness of the agreement on passage of such legislation by Congress and also establish a basis for compensation for takings under the Fifth Amendment should a subsequent Congress decide to take action contrary to that commitment. For instance, a later Congress has the power to set aside the actions of an earlier Congress, but the question of takings and just compensation then arises. If these conditions are not incorporated, the States making the agreement may later find that federal actions render their agreement ineffective.

Cross-references: §2A-1-03 (duration of agreement); §2A-1-04 (amendments and supplements); §2A-2-05 (party or parties).

Similar Agreements: *Republican River Compact*, 57 Stat. 86 (1943); *Belle Fourche River Compact*, 58 Stat. 94 (1944); *Delaware River Basin Compact*, Pub. L. 87-328, 75 Stat. 688 (1961); *Susquehanna River Basin Compact*, Pub. L. No. 91-575, 84 Stat. 1509 (1970); *Convention on the Protection and Use of Transboundary Watercourses and International Lakes*,

31 I.L.M. 1312 (1992); *Convention on the Law of the Non-Navigational Uses of International Watercourses*, United Nations Document A/51/869 (1998).

§2A-1-02 CONSENT TO JURISDICTION (Optional, for U.S. use)

This agreement shall be effective once the United States Congress gives its consent for the United States to be named and joined as a Party defendant or otherwise in any case or controversy involving the construction or application of this agreement in which one or more of the States is a plaintiff, without regard to any requirement as to the sum or value in controversy or diversity of citizenship of Parties to the case or controversy.

Commentary: The predominance of federal interests in water resources makes it likely that any litigation concerning the agreement between States will involve federal interests. The doctrine of sovereign immunity could prevent joinder of the federal interests as Parties to the suit absent a waiver of sovereign immunity. The discretionary devision not to join federal parties led to dismissal of a suit filed by Texas against New Mexico in 1951 to enforce certain provisions of the *Rio Grande Compact of 1938*, 53 Stat. 785, 938. The Supreme Court dismissed the case because the federal government was not joined as a party, but had important interests that would be affected by any such suit. *Texas v. New Mexico*, 352 U.S. 991, 957. In 1952, Congress enacted the *McCarren Amendment,* 43 U.S.C. 66, which waived federal sovereign immunity to be joined in general stream adjudications. However, in most cases, the agreement will cover management and development issues that reach beyond general stream adjudication. The Parties should consider including such a waiver of sovereign immunity as a condition to effectiveness of the agreement. They may also wish to add a provision granting jurisdiction over any such cases to the District Courts, which may be preferable to the Supreme Court as the initial forum for resolving certain types of disputes. The *Red River Compact*, 94 Stat. 3305 (1980) takes this approach.

Cross References: §1A-1-04 (preservation of federal right); §1A-1-05 (national security); §2A-2-05 (party or parties).

Similar Agreements: *Kansas-Nebraska Big Blue River Compact*, 86 Stat. 193 (1972); *Red River Compact*, 94 Stat. 3305 (1980).

§2A-1-03 DURATION OF AGREEMENT (Optional)

(a) The Parties intend that the duration of this Agreement shall be for an initial period of (_) years from its effective date. Notification of the withdrawal must be made (__) months in advance of the prospective withdrawal. If none of the signatory Parties notifies the Commission of intention to terminate the Agreement at the end of the then current (__) year period, the Agreement shall be continued for an additional period of (__) years.

(b) In the event that this Agreement should be terminated by operation of paragraph (a) above, the management structure for the Agreement shall be dissolved, its assets and liabilities transferred equitably to the Parties, and its corporate affairs completed in such manner as may be provided by agreement of the signatory Parties.

Commentary: The Parties may prefer to establish no duration and rely on later provisions to modify or terminate the agreement. However, two significant principles are established by this provision. First, setting a duration for an extended period of time, recommended at 50 years, allows for predictability on terms of water resources development; it also allows sufficient time to recover capital costs in the financing of projects. Second, establishing a duration ensures that the Parties reconsider the Agreement only after a sufficient hydrologic record is established. It should be noted, however, that this provision greatly impacts on the exercise of sovereignty of the Parties involved.

Cross-references: §2A-1-01 (effective date); §2A-1-04 (amendments and supplements); §2A-1-05 (powers of sovereign parties; withdrawals).

Similar Agreements: *Delaware River Basin Compact* (DRBC), Pub. L. 87-328, 75 Stat. 688 (1961); *Susquehanna River Basin Compact*, Pub. L. No. 91-575, 84 Stat. 1509 (1970).

§2A-1-04 POWERS OF SOVEREIGN PARTIES; WITHDRAWAL
(Optional, international only)

Nothing in this Agreement shall be construed to relinquish the functions, powers or duties of the government of any signatory Party with respect to the control of any waters within its territory, nor shall any provision hereof be construed in derogation of any of the powers of the Parties to regulate commerce within their sovereign borders.

Commentary. This provision acknowledges the inherent sovereignty of the individual Parties and recognizes that any relinquishment of sovereignty is limited solely to the purposes of this Agreement.

Cross-references: §1A-1-02 (coordination and cooperation); §1A-1-03 (good faith implementation); §2A-2-06 (waters of the basin); §4A-1-01 (exclusive jurisdiction and control).

Similar Agreements: *Delaware River Basin Compact*, Pub. L. 87-328, 75 Stat. 688 (1961); *Susquehanna River Basin Compact*, Pub. L. No. 91-575, 84 Stat. 1509 (1970); *Apalachicola-Chattahoochee-Flint River Basin Compact*, O.C.G.A. 12-10-100 (1997); *Alabama-Coosa-Tallapoosa River Basin Compact*, O.C.G.A. 12-10-110 (1997); *Convention on the Protection and Use of Transboundary Watercourses and International Lakes*, 31 I.L.M. 1312 (1992); *The North American Agreement on Environmental Cooperation between the Government of the United States of America, the Government of Canada, and the Government of the United*

Mexican States, 32 I.L.M. 1480 (1993); *Agreement on Cooperation for the Sustainable Development of the Mekong River Basin*, 34 ILM 864 (1995).

§2A-1-05 AMENDMENTS AND SUPPLEMENTS (Optional)

The provisions of this agreement shall remain in full force and effect until amended by action of the governing bodies of each of the Parties and consented to and approved by any other necessary authority in the same manner as this agreement is initially ratified.

Commentary. Agreements may, over time, fail to operate as well as initially intended. Therefore, some amendment process should be specified. In some cases, the approval of another institution may be required. If, for example, the agreement is between states of the United States, the United States Constitution arguably requires Congressional approval of any amendment as well as approval of the original agreement, unless the agreement provides for a different method of amendment. In this latter case, the Congressional approval of the initial agreement would implicitly grant consent to modify the agreement in accordance with the terms of the agreement. If the agreement is between sovereign nations, the references to other "necessary authority" may be omitted, but the particular circumstances of each case must be considered.

Cross-references: §2A-1-07 (limited applicability).

Similar Agreements: *Delaware River Basin Compact*, Pub. L. 87-328, 75 Stat. 688 (1961); *Susquehanna River Basin Compact*, Pub. L. No. 91-575, 84 Stat. 1509 (1970); *Convention on the Protection and Use of Transboundary Watercourses and International Lakes*, 31 I.L.M. 1312 (1992); *Agreement on Cooperation for the Sustainable Development of the Mekong River Basin*, 34 ILM 864 (1995).

§2A-1-06 EXISTING AGENCIES (Optional)

It is the purpose of the signatory Parties to preserve and utilize the functions, powers and duties of existing offices and agencies of the individual governments to the extent not inconsistent with the Agreement, and the institution established to enforce this Agreement is authorized and directed to utilize and employ such offices and agencies for the purpose of this Agreement to the fullest extent it finds feasible and advantageous.

Commentary: The use of existing offices and agencies prevents duplication, and consequently reduces the costs, of data collection and management of the water resource subject to the Agreement.

Cross-references: §3A-1-01 (alternative 1, use of party officials); §3A-1-05 (alternative 2, commission organization and staffing); §3A-1-06 (alternative 2, advisory boards); §3A-1-07 (alternative 2, rules of procedures); §3A-2-01 (alternative 2, general powers and duties).

Similar Agreements: *Delaware River Basin Compact*, Pub. L. 87-328, 75 Stat. 688 (1961); *Susquehanna River Basin Compact*, Pub. L. No. 91-575, 84 Stat. 1509 (1970); *Convention on the Protection and Use of Transboundary Watercourses and International Lakes*, 31 I.L.M. 1312 (1992).

§2A-1-07 SEVERABILITY

Should a tribunal of competent jurisdiction hold any part of this Agreement to be void or unenforceable, it shall be considered severable from those portions of the greement capable of continued implementation in the absence of the voided provisions. All other severable provision capable of continued implementation shall continue in full force and effect.

Commentary: The drafters of the Agreement should consider whether they wish this clause to be included. The advantage of such a clause is that it avoids the possibility of having the entire Agreement become null and void if any part is found to be void or unenforceable.

Cross-references: §2A-1-04 (amendments and supplements).

Similar Agreements: *Yellowstone River Compact,* 65 Stat. 663 (1950); *Sabine River Compact*, 68 Stat. 690 (1953); *Klamath River Basin Compact*, 71 Stat. 497 (1957); *Delaware River Basin Compact*, Pub. L. 87-328, 75 Stat. 688 (1961); *Susquehanna River Basin Compact*, Pub. L. No. 91-575, 84 Stat. 1509 (1970).

§2A-1-08 ANNEXES

The Annexes to this Agreement, to the extent that they are consistent with the objectives and intent of the Agreement, constitute an integral part of the Agreement.

Commentary: An effective water management agreement will necessarily contain detailed information and data of a procedural nature. While such information may be essential to the effectiveness of the particular agreement, its inclusion in the main body of the agreement may take away from the essence of the contractual nature of the agreement. The use of annexes minimizes this effect.

Cross-references: Annex A (arbitral panel).

Similar Agreements: ASEAN Agreement on the Conservation of Nature and Natural Resources, 1985; *The North American Agreement on Environmental Cooperation between the Government of the United States of America, the Government of Canada, and the Government of the United Mexican States*, 32 I.L.M. 1480 (1993).

Part 2 Definitions

§2B-2-01 _____ BASIN

_____ Basin" means the area of drainage into the _____ River and its tributaries, [and] aquifers underlying the drainage, or only the aquifers themselves..

Commentary: The Agreement could include the total surface area of drainage throughout the Basin and contain aquifers underlying the surface drainage. Some tributaries can be connected to the underlying aquifers holding the underground water. Some of the aquifers could be connected to more than one of the surface water basins. The geographic scope of the Agreement should be defined to ensure there are no future disagreements about what lands are or are not covered by the Agreement. A map may be incorporated, but care should be taken that the map is cartographically accurate. Because the map is likely to be at a scale too small for precise delineation of boundaries, it should be made clear that it is for general reference only. In the event of a dispute over land or within the defined _____ River and its tributaries, the actual limits of the watershed as determined on the ground should be controlling

Cross References: §1A-1-01 (general policies); §1A-1-04 (preservation of federal rights).

§2A-2-02   DROUGHT

"Drought" conditions means conditions brought about by the lack of precipitation or water stored in the soil, in a quantity agreeable to the Parties, from the mean annual [rainfall] precipitation and water measured in soil.

Commentary: Management action will arise from a drought, or lack of mean annual rainfall, but could arise from other causes as well, such as the collapse of a dam with the resulting draining of a reservoir on which the Commission users depend. Mather (1974) indicates the number of quite different definitions is large. The definition should be determined, in large measure, by the use intended. Then a "drought management strategy" would be a specific course of conduct planned by the Commission as a necessary or appropriate response to the lack of precipitation.

Cross-references: §3A-2-01 (general powers and duties).

§2A-2-03   FLOOD

"Flood" conditions means conditions resulting from heavy runoff with a frequency agreeable to the Parties.

Commentary: The flood condition is almost the opposite of a drought. A large amount of water is to be controlled by facilities of the Commission. The Parties are to agree as to the frequency of the flow of high waters in the Basin. Most of the time, these flows are during periods that exceed the amount of flow that occurs during the years of mean annual precipitation.

Cross-references: §3A-2-01 (general powers and duties); §4A-1-02 (data exchange).

§2A-2-04 INTERBASIN TRANSFER

An "interbasin transfer" is any transfer of water, in excess of (____) gallons/liters per day, from one water basin to another.

Commentary: This definition merely makes clear that the term "interbasin transfer" is not limited in any fashion but refers to all transfers from one water basin to another. The provisions regarding interbasin transfers allow regulations to exempt certain small transfers. In many states within the United States, 100,000 gallons per day (378,571 liters per day) are exempt from regulation.

Cross-references: §4A-1-02 (data exchange).

§2A-2-05 PARTY OR PARTIES

"Party or Parties" means, unless the text otherwise indicates, those governmental units signatory to this Agreement.

Commentary: Defining the terms in this way avoids the need to include similar language at numerous points throughout the agreement. As a matter of law, it may be unnecessary to state this principle.

Cross References: §1A-1-02 (coordination and cooperation); §1A-1-03 (good faith implementation); §1A-1-05; §2A-1-01 (effective date); §2A-1-02 consent to jurisdiction; §2A-1-05 (powers of sovereign parties); §3A-1-01 (alternative 1, use of party officials); §3A-1-02 (status, immunities and privileges); §3A-1-05 (commission organization and staffing); §3A-1-06 (advisory boards); §3A-2-01 (general powers and duties; §3A-2-02 (special powers and duties); §4A-1-01 (exclusive jurisdiction and control); §4A-1-02 (data exchange); Article 5A (dispute resolution).

§ 2A-2-06 WATERS OF THE BASIN

"Waters of the Basin" shall include all water found within the Basin, whether surface, underground, or atmospheric water.

Commentary: This definition should be included to make it clear that underground water and atmospheric water are included within the scope of the agreement, if that is the intent of the Parties. The technological questions relating to atmospheric water may result in uncertainty regarding its allocation, but to the extent the Parties wish to reach a complete agreement, the matter should be addressed, or recognition given to the fact that the Parties have chosen to reserve that issue for later resolution. The Parties should also decide whether water imported from other basins should be included within the scope of the agreement. If it is not to be so included, that exclusion should be noted in this section.

Cross References: §2A-1-05 (powers of sovereign parties; withdrawal).

ARTICLE 3A

ADMINISTRATION

Part 1 Administrative Authority

<u>Alternative 1</u> **(administration by party officials)**

§3A-1-01 USE OF WATER MANAGEMENT OFFICIALS OF THE PARTIES

It shall be the duty of the Parties to administer this Agreement through the official of each Party who is now or may hereafter be charged with the duty of administering the public water resources management, and to collect and correlate through such officials the data necessary for the proper administration of the provisions of this Agreement. Such officials may, by unanimous action, adopt rules and regulations consistent with the provisions of this agreement.

Commentary: This alternative provides an austere means of administering the agreement. It conceives of no additional administrative apparatus beyond the administrative machinery that exists with the Parties.

Cross-reference: §1A-1-01 (general policies); §1A-1-02 (coordination and cooperation); §1A-1-03 (good faith implementation); §2A-2-05 (party or parties); §3A-1-02 (alternative 1, substitution of officials); §3A-1-03 (altrnative 1, implementation and verification of agreement); §3A-1-04 (alternative 1, funding).

Similar Agreements: *La Plata River Compact*, 43 Stat. 796 (1925); *Republican River Compact*, 57 Stat. 86 (1943); *Snake River Compact*, 64 Stat. 29 (1949); *Costilla Creek Compact*, 60 Stat. 246 (1946); amended 77 Stat. 350 (1963); *Upper Niobrara River Compact*, 83 Stat. 86 (1969).

§3A-1-02 SUBSTITUTION OF OFFICIALS

Whenever any official of any Party is designated to perform any duty under this agreement, such designation shall be interpreted to include the Party's official or officials upon whom the duties now performed by such official may hereafter devolve.

Commentary: This section is included to guard against confusion in the event there is a subsequent reorganization of a party's government.

Cross References: §1A-1-01 (general policies); §1A-1-02 (coordination and cooperation); §1A-1-03 (good faith implementation); §2A-2-05 (party or parties); §3A-1-01 (use of party officials, alternative 1); §3A-1-03 (implementation and verification of agreement, alternative 1); §3A-1-04 (funding, alternative 1).

Similar Agreements: *La Plata River Compact*, 43 Stat. 796 (1925); *South Platte River Compact*, 44 Stat. 195 (1923).

§3A-1-03 IMPLEMENTATION AND VERIFICATION OF AGREEMENT

(a) Each Party shall identify or maintain the administrative machinery necessary to implement the provisions of this Agreement, and, where several governmental institutions are involved, create the necessary co-ordinating mechanism for the authorities dealing with designated aspects of the Agreement.

(b) Each Party shall have the duty to establish, maintain, and operate such suitable water gaging stations and facilities for measuring water quantity and quality as it finds necessary to administer and effect verification of this agreement. Reliance on any existing facility operated by a regional authority, national government or international organization shall not relieve the Party from ensuring that data of sufficient quality and quantity is available to administer and verify this agreement.

Commentary: Implementation and verification of the Agreement will require administrative and technical support that must be provided by the Parties. This section obligates the Parties to provide that support.

Cross-references: §1A-1-01 (general policies); §1A-1-02 (coordination and cooperation); §1A-1-02 (good faith implementation); §3A-1-01 (alternative 1, use of party officials); §3A-1-04 (alternative 1, funding).

Similar Agreements: *La Plata River Compact*, 43 Stat. 796 (1925).

§3A-1-04 FUNDING

Each Party shall allocate sufficient qualified personnel with adequate enforcement powers and sufficient funds to accomplish the tasks necessary for the implementation of this Agreement.

Commentary: In the case of simple allocation agreements in which no commission is established, funding provisions are not normally included. However, in order to ensure no misunderstanding exists concerning the responsibilities of each Party, an explicit provision may be preferable. Section 3B-1-04 is designed to avoid disputes over financing by requiring that each Party will operate the necessary facilities within its borders.

Cross References: §1A-1-01 (general policies); §1A-1-02 (coordination and cooperation); §1A-1-02 (good faith implementation); §3A-1-01 (alternative 1, use of party officials); §3A-1-02 (alternative 1, substitution of officials); §3A-1-03 (altrnative 1, implementation and verification of agreement).

Similar Agreements: ASEAN Agreement on the Conservation of Nature and Natural Resources, 1985.

Alternative 2 (administration by commission)

§3A-1-01 COMMISSION CREATED

(a) The _____ Commission (hereinafter the Commission) is hereby created as a body politic and corporate, with succession for the duration of this Agreement, as an agency and instrumentality of the governments of the respective signatory Parties.

(b) The Commission shall develop and effectuate policies for cooperation, coordination and provision of information, concerning all planning and management activities and water projects affecting their common water resources, in order to assist the Parties in sharing their shared water resources in an equitable and reasonable manner.

Commentary: The name of the Commission should reflect the geographical setting of the particular water resources, usually a river designating the border between the Parties or a river running across their common border, to which the Agreement refers.

To be effective in coordinating water utilization and precluding conflict, the institutional structure must necessarily stimulate cooperation between or among governmental institutions of the signatory Parties. It should also be able to (1) determine the facts of water use in the territory of each Party, (2) resolve disputes across the boundaries between the Parties, (3) guide responses to unusual temporary water emergencies, (4) regulate or design long-term solutions, and (5) enforce its decision. *See* e.g. U.N. Department of Economic and Social Affairs, Natural Resources Water Series No. 1, *Management of Int'l Water Resources, Institutional and Legal Aspects,* Report of the Panel of Experts on the Legal and Institutional Aspects of International Water Resources Development, U.N. Doc. St/ESA/5; *See* also David Le Marquand, *Politics of International Basin Cooperation and Management,* in Water in a Developing World [Albert E. Utton & Ludwick A. Teclaff, eds. (1978)].

The organizational structure of the Commission should be constituted according to the specifics of the water resource itself and the political structures of the parties involved. What works for wealthy nations may not work for developing countries, and, in some shared water situations, cultural differences (e.g., the Jordan River) may require different management structures than used in those situations involving similar cultures (e.g., the Rhine River). *See* Arthur E. Williams, *The Role of Technology in Sustainable Development* in Water Resources Administration in the United States [M. Reuss, ed. (1993)]. Sterner states that "there must be an appropriate legal structure and set of institutions that define property rights and establish the framework within which an environmental authority can function." *See* Thomas Sterner, Economic Policies for Sustainable Development, [Thomas Sterner, ed. (1994)].

Cross-references: §1A-1-01 (general policies); 3-1-02 (alternative 2, commission jurisdiction); §3A-1-03 (alternative 2, commissioners); §3A-1-04 (alternative 2, status, immunities and privileges,); §3A-1-05 (alternative 2, commission organization and staffing); §3A-1-06 (alternative 2, advisory boards); §3A-1-07 (rules of procedures); §3A-1-02 (alternative 2, commission jurisdiction); §3A-1-03 (alternative 2, commissioners); §3A-1-04 (alternative 2, status, immunities and privileges); §3A-1-05 (alternative 2, commission organization and staffing); §3A-1-06 (alternative 2, advisory boards); §3A-1-07 (alternative 2, rules of procedures); §3A-2-01 (alternative 2, general powers and duties); §3A-2-02 (alternative 2, special powers and duties); §3A-3-01 (alternative 2, meetings, hearings and records); §3A-3-02 (alternative 2, funding and expenses of the commission; §4A-1-01 (exclusive jurisdiction and control); §4A-1-02 (data exchange); Article 5A (dispute resolution).

Similar Agreements: *Klamath River Basin Compact,* 71 Stat. 497 (1957); *Delaware River Basin Compact,* Pub. L. 87-328, 75 Stat. 688 (1961); *Susquehanna River Basin Compact,* Pub. L. No. 91-575, 84 Stat. 1509 (1970).; *Apalachicola-Chattahoochee-Flint River Basin Compact,* O.C.G.A. 12-10-100 (1997); *Alabama-Coosa-Tallapoosa River Basin Compact,*O.C.G.A. 12-10-110 (1997); *Treaty between the United States and Great Britain relating to Boundary Waters, and Questions arising between the United States and Canada,* 36 Stat. 2451 (1909); *Treaty between the United States of America and Mexico, Utilization of Waters of the Colorado and Tijuana rivers and of the Rio Grande,* 59 Stat. 1219 (1945); *Indus Water Treaty,* 419 UNTS 126 (1960); *The North American Agreement on Environmental Cooperation between the Government of the United States of America, the Government of Canada, and the Government of the United Mexican States,* 32 I.L.M. 1480 (1993); *Agreement on Cooperation for the Sustainable Development of the Mekong River Basin,* 34 ILM 864 (1995).

§3A-1-02   COMMISSION   AUTHORITIES AND RESPONSIBILITIES

The Commission shall coordinate all programs of data collection and dissemination within the _____ River Basin and shall facilitate the resolution of any controversy or conflict involving the use, obstruction or diversion of those waters within the Basin that affects water availability and quality.

Commentary.   This provision describes the geographic and hydrologic authorities and responsibilities of the Commission.   Similar provisions appear in all agreements in order to clarify and define the limits of authority ceded by the Parties to an independent administrative unit.   *See*   Hayton, R.D. and A.E. Utton, Transboundary Groundwaters: The Bellagio Draft Treaty, 1989.

Cross-references: §1A-1-01 (general policies); §2A-1-02 (consent to jurisdiction); §3A-1-01 (alternative 2, commission created); §3A-1-03 (alternative 2, commissioners); §3A-1-04 (alternative 2, status, immunities and privileges); §3A-1-05 (alternative 2, commission organization and staffing); §3A-1-06 (alternative 2, advisory boards); §3A-1-07 (alternative 2, rules of procedures); §3A-2-01 (alternative 2, general powers and duties); §3A-2-02 (alternative 2, special powers and duties); §3A-3-01 (alternative 2, meetings, hearings and records§3A-3-02 (alternative 2, funding and expenses of the commission; §4A-1-01 (exclusive jurisdiction and control); §4A-1-02 (data exchange); Article 5A (dispute resolution).

Similar Agreements: *Delaware River Basin Compact,* Pub. L. 87-328, 75 Stat. 688 (1961); *Susquehanna River Basin Compact,* Pub. L. No. 91-575, 84 Stat. 1509 (1970); *Treaty between the United States and Great Britain relating to Boundary Waters, and Questions arising between the United States and Canada,* 36 Stat. 2451 (1909); Treaty Respecting Utilization of Water in the Colorado and Tijuana Rivers and of the Rio Grand, 59 Stat. 1219 (1945); *Agreement on Cooperation for the Sustainable Development of the Mekong River Basin,* 34 ILM 864 (1995).

§3A-1-03   COMMISSIONERS

Alternative 1

(a) The Commission shall be governed by a Board of Commissioners, consisting of the Governor or other Chief Executive Officer of each of the signatory Parties plus a federal representative appointed by the President of the United States.

(b) Each Commissioner may appoint an alternate to act in his place and stead, with authority to attend all meetings of the Commission, and with power to vote in the absence of the Commissioner. Unless otherwise provided by law of the signatory Party for which he is appointed, each alternate shall serve during the term of the Commissioner appointing him, subject to removal at the pleasure of the Commissioner. In the event of a vacancy in the office of alternate, it shall be filled in the same manner as an original appointment for the unexpired term only.

(c) All matters requiring a policy decision affecting the substance of the Agreement shall be decided upon by the Commissioners and unanimity shall be required. For all other matters, the rule of decision shall be simple majority. Each member shall be entitled to one vote on all matters that may come before the Commission. No action by the Commissioners shall be taken at any meeting unless a majority of the membership shall vote in favor thereof. The federal representative shall have no vote.

Alternative 2

(a) The signatory Parties agree to establish and maintain the _____ Commission composed of (___) commissioners, (___) on the part of each sovereign Party, appointed by the (President or Prime Minister or Head of State) thereof.

(b) The Chair of the Commission shall be for a term of (___) years and rotate according to the alphabetical listing of the signatory parties.

Commentary. Alternative 1 recognizes that the responsibility (and authority) for all policy decisions affecting the substance of the Agreement remains with the principal executive office of the respective Parties. The membership of the typical commission in the U.S. includes one or more members from each Party plus a federal representative. However, in most instances, the federal representative has no vote. *See* Grant, §46.03. The *Delaware River Basin Compact* and *Susquehanna River Basin Compact* are notable exceptions.

Alternative 2 is presented under the assumption that two Parties are involved. When more than two Parties are involved, appropriate changes to the composition of the Commission should be made. Alternative 2 is less specific and consequently may be more appropriate for the international context.

Cross-references: §1A-1-01 (general policies); §1A-1-04 (preservation of federal rights); §1A-1-05 (national security); §2A-1-02 (consent to jurisdiction); §2A-1-04 (amendments and supplements); §2A-1-05 (powers of sovereign parties; withdrawal); §2A-1-06 (existing agencies); §2A-1-07 (limited applicability); §2A-2-05 (party or parties); §3A-1-01 (alternative 2, commission created); §3A-1-02 (alternative 2, commission jurisdiction); §4A-1-01 (exclusive jurisdiction and control).

Similar Agreements: *Klamath River Basin Compact*, 1957; *Delaware River Basin Compact*, Pub. L. 87-328, 75 Stat. 688 (1961); *Susquehanna River Basin Compact*, Pub. L. No. 91-575, 84 Stat. 1509 (1970); *Treaty between the United States and Great Britain relating to Boundary Waters*, 36 Stat. 2451 (1909); *Treaty between the United States of America and Mexico, Utilization of Waters of the Colorado and Tijuana rivers and of the Rio Grande*, 59 Stat. 1219 (1945); Convention Between the Federal Republic of Germany and the Czech and Slovak Federal Republic and the European Economic Community on the International Commission for the Protection of the Elbe, International Environmental Law, Multilateral Agreements, 976:90/1 (1990); *Convention on the Protection and Use of Transboundary Watercourses and International Lakes*, 31 I.L.M. 1312 (1992); *The North American Agreement on Environmental Cooperation between the Government of the United States of America, the Government of*

Canada, and the Government of the United Mexican States, 32 I.L.M. 1480 (1993); *Agreement on Cooperation for the Sustainable Development of the Mekong River Basin*, 34 ILM 864 (1995).

§3A-1-04 STATUS, IMMUNITIES AND PRIVILEGES (Optional)

To enable the Commission to fulfill its purpose and the functions with which it is entrusted, the status, immunities and privileges set forth in this Article shall be accorded to the Commission in the territories of each Party.

(a) The Commission, its property and its assets, wherever located, and by whomsoever held, shall enjoy the same immunity from suit and every form of judicial process as is enjoyed by the Parties, except to the extent that the Commission may expressly waive its immunity for the purposes of any proceedings or by the terms of any contract.

(b) Property and assets of the Commission, wheresoever located and by whomsoever held, shall be considered public property and shall be immune from search, requisition, confiscation, expropriation or any other form of taking or foreclosure by executive or legislative action.

(c) To the extent necessary to carry out the purpose and functions of the Commission and to conduct its operations in accordance with this Agreement, all property and other assets of the Commission shall be free from restrictions, regulations, controls and moratoria of any nature affecting the implementation of this agreement, except as may otherwise be provided in this Agreement.

(d) The official communications of the Commission shall be accorded by each Party the same treatment that it accords to the official communications of the other Parties.

(e) (Optional, for international use only) The Commissioners and other personnel engaged directly in the affairs of the Commission shall have the following privileges and immunities:

(1) Immunity from legal process with respect to acts performed by them in their official capacity except when the Commission expressly waives this immunity.

(2) When not citizens of one of the signatory Parties, the same immunities from immigration restrictions, alien registration requirements and national service obligations and the same facilities as regards exchange provisions as are accorded by each Party to the representatives, officials, and employees of comparable rank of the other Party; and

(3) The same privileges in respect of traveling and facilities as are accorded by each Party to representatives, officials, and employees of comparable rank of the other Party.

(f) The Commission, its property, other assets, income, and the operations it carries out pursuant to this Chapter shall be immune from all state taxation. The Commission shall also be immune from any obligation relating to the payment, withholding or collection of any tax or customs duty. No state tax shall be levied on or in respect of salaries and benefits paid by the Commission to officers or staff of the Commission who are not local citizens.

(g) Each Party, in accordance with its juridical system, shall take such action as is necessary to make effective in its own territories the principles set forth in this Article, and shall inform the Commission of the action which it has taken on the matter.

Commentary: This provision provides the Commissioners and their personnel with the same legal protections that normally exist for governmental officials of the Parties.

Cross-references: §1A-1-01 (general); §2A-2-05 (party or parties); §3A-1-01 (alternative 2, commission created); §3A-1-02 (alternative 2, commission jurisdiction); §3A-1-03 (alternative 2, commissioners); §3A-1-05 (alternative 2, commission organization and staffing); §3A-1-06 (alternative 2, advisory boards); §3A-1-07 (alternative 2, rules of procedures); §3A-2-01 (alternative 2, general powers and duties); §3A-2-02 (alternative 2, special powers and duties); §3A-3-01 (alternative 2, meetings, hearings and records); §3A-3-02 (alternative 2, funding and expenses of the commission); §4A-1-01 (exclusive jurisdiction and control); §4A-1-02 (data exchange).

Similar Agreements: *Agreement between the Governments of the United States of America and the Government of the United Mexican States Concerning the Establishment of a Border Environment Cooperation Commission and the North American Development Bank*, 19 U.S.C. §473 (1993).

§3A-1-05 COMMISSION ORGANIZATION AND STAFFING

(a) The Commission shall meet and organize as promptly after the members thereof are appointed, and when organized the Commission may fix such times and places for its meetings as may be necessary, subject at all times to special call or direction by agreement of the Parties. Each Commissioner, upon the first joint meeting of the Commission after his appointment, shall, before proceeding with the work of the Commission, make and subscribe a solemn declaration in writing that he will faithfully and impartially perform the duties imposed upon him under this Agreement, and such declaration shall be entered on the records of the proceedings of the Commission.

(b) The respective Commissioners may each appoint a secretary, and these shall act as joint secretaries of the Commission at its joint sessions, and the Commissioners may utilize professional and administrative personnel from existing agencies of the respective Parties from time to time, as it may deem advisable. The salaries and personal expenses of the Commissioners and their supporting staff shall be paid by their respective Governments.

Commentary: This provision provides the authority and instructions for the organization and initiation of Commission undertakings.

Cross-references: §1A-1-01, (general policies); §1A-1-02, Purposes of Agreement; §1A-1-03, Objectives of Agreement; §2A-2-05 (party or parties); §3A-1-01 (alternative 2, commission created); §3A-1-02 (alternative 2, commission jurisdiction); §3A-1-03 (alternative 2, commissioners); §3A-1-04 (alternative 2, status, immunities and privileges); §3A-1-06 (alternative 2, advisory boards); §3A-1-07 (alternative 2, rules of procedures); §3A-2-01 (alternative 2, general powers and duties); §3A-2-02 (alternative 2, special powers and duties); §3A-3-01 (alternative 2, meetings, hearings and records).

Similar Agreements: *Treaty between the United States and Great Britain relating to Boundary Waters*, 36 Stat. 2451 (1909); Convention Between the Federal Republic of Germany and the Czech and Slovak Federal Republic and the European Economic Community on the International Commission for the Protection of the Elbe, International Environmental Law, Multilateral Agreements, 976:90/1 (1990).

§3A-1-06 ADVISORY BOARDS (Optional)

(a) The Commission may appoint an advisory board or boards, composed of qualified persons to conduct on its behalf investigations and studies that may be necessary or desirable and to report to the Commission regarding any questions or matters involved in the subject matter of the reference.

(b) Such boards ordinarily will have an equal number of members from each Party.

(c) The Commission will make copies of the main or final report of such board or a digest thereof available for examination by all Parties.

Commentary: This provision acknowledges that the resolution of some disputes involving water may require specialists and experts with particular knowledge or talent not available within the Commission structure.

Cross-references: §1A-1-01 (general policies); §2A-2-05 (party or parties); §3A-1-01 (alternative 2, commission created); §3A-1-02 (alternative 2, commission jurisdiction); §3A-1-03 (alternative 2, commissioners); §3A-1-04 (alternative 2, status, immunities and privileges); §3A-1-05 (alternative 2, commission organization and staffing); §3A-1-07 (alternative 2, rules of procedures); §3A-2-01 (alternative 2, general powers and duties); §3A-2-02 (alternative 2, special powers and duties); §3A-3-01 (alternative 2, meetings, hearings and records); §4A-1-01 (exclusive jurisdiction and control); §4A-1-02 (data exchange).

Similar Agreements: *Treaty between the United States and Great Britain relating to Boundary Waters*, 36 Stat. 2451 (1909).

§3A-1-07 RULES OF PROCEDURE

The Commission shall adopt its own Rules of Procedure, and may seek technical advisory services, as it deems necessary.

Commentary: Some compacts contain detailed rules of procedures. Such detail can, however, cause unnecessary complications to effective, efficient and timely response to emergencies and extreme hydrological conditions such as floods or droughts. Except when necessary for policy reasons, the Agreement should not bind the Commission to specific procedural requirements.

Cross-references: §3A-1-01 (alternative 2, commission created); §3A-1-02 (alternative 2, commission jurisdiction); §3A-1-03 (alternative 2, commissioners); §3A-1-04 (alternative 2, status, immunities and privileges); §3A-1-05 (alternative 2, commission organization and staffing); §3A-1-06 (alternative 2, advisory boards); §3A-2-01 (alternative 2, general powers and duties); §3A-2-02 (alternative 2, special powers and duties); §3A-3-01 (alternative 2, meetings, hearings and records).

Similar Agreements: *Klamath River Basin Compact*, 71 Stat. 497 (1957); *Delaware River Basin Compact*, Pub. L. 87-328, 75 Stat. 688 (1961); *Susquehanna River Basin Compact*, Pub. L. No. 91-575, 84 Stat. 1509 (1970); *The North American Agreement on Environmental Cooperation between the Government of the United States of America, the Government of Canada, and the Government of the United Mexican States*, 32 I.L.M. 1480 (1993); *Agreement on Cooperation for the Sustainable Development of the Mekong River Basin*, 34 ILM 864 (1995); *Convention on the Law of the Non-Navigational Uses of International Watercourses*, United Nations Document A/51/869 (1998).

Part 2 Powers and duties

§3A-2-01 GENERAL POWERS AND DUTIES

The Commission shall have the power and responsibility

(a) To adopt bylaws and procedures governing its conduct;

(b) To sue and be sued in any court of competent jurisdiction;

(c) To retain and discharge professional, technical, clerical and other staff and such consultants as are necessary to accomplish the purposes of this Agreement;

(d) To receive funds from any lawful source and expend funds for any lawful purpose;

(e) To enter into agreements or contracts, where appropriate, in order to accomplish the purposes of this Agreement;

(f) To create committees and delegate responsibilities;

(g) To plan, coordinate, monitor, and make recommendations for the use of the water resources shared by the Parties for the purposes of, but not limited to, coordination and cooperation for the purposes of minimizing adverse impacts of floods and droughts, and facilitating the utilization of the waters as may be deemed appropriate by the Commission;

(h) To participate with other governmental and non-governmental entities in carrying out the purposes of this Agreement;

(i) To perform all functions required of it by this Agreement and to do all things necessary, proper or convenient in the performance of its duties hereunder, either independently or in cooperation with any Party.

Commentary: Certain authorities and responsibilities pertain to any organizational structure, whether the structure be a business corporation, a governmental agency or an institution developed to manage international relations. This list provides the Commission with the minimum powers necessary to achieve the purposes assigned to it.

Cross-references: §1A-1-04 (preservation of federal rights); §1A-1-05 (national security); §2A-1-02 (consent to jurisdiction); §2A-1-03 (duration of agreement); §2A-1-04 (amendments and supplements); §2A-1-05 (powers of sovereign parties; withdrawal); §2A-1-06 (existing agencies); §2A-1-07 (limited applicability); §2A-2-02 (drought); §2A-2-03 (flood); §2A-2-05 (party or parties); §3A-1-07 (alternative 2, rules of procedures); §3A-2-02 (alternative 2, special powers and duties); §3A-3-01 (alternative 2, meetings, hearings and records§3A-3-02 (alternative 2, funding and expenses of the commission); §4A-1-01 (exclusive jurisdiction and control); §4A-1-02 (data exchange); Article 5A (dispute resolution).

Similar Agreements: *Apalachicola-Chattahoochee-Flint River Basin Compact*, O.C.G.A. 12-10-100 (1997); *Alabama-Coosa-Tallapoosa River Basin Compact*, 1997; *Delaware River Basin Compact*, Pub. L. 87-328, 75 Stat. 688 (1961); *Susquehanna River Basin Compact*, Pub. L. No. 91-575, 84 Stat. 1509 (1970); *Klamath River Basin Compact*, 1957; *Treaty between the United States of America and Mexico, Utilization of Waters of the Colorado and Tijuana rivers and of the Rio Grande*, 59 Stat. 1219 (1945); Convention Between the Federal Republic of Germany and the Czech and Slovak Federal Republic and the European Economic Community on the International Commission for the Protection of the Elbe, International Environmental Law,

Multilateral Agreements, 976:90/1 (1990); *The North American Agreement on Environmental Cooperation between the Government of the United States of America, the Government of Canada, and the Government of the United Mexican States*, 32 I.L.M. 1480 (1993); *Agreement on Cooperation for the Sustainable Development of the Mekong River Basin*, 34 ILM 864 (1995).

§3A-2-02 SPECIAL POWERS AND DUTIES (Optional)

(a) The Commissioners shall develop and implement a comprehensive system to facilitate the exchange of information between the Parties. This system of information exchange shall be sufficient in scope to adequately promote the utilization of the water resources shared by the Parties.

(b) The Parties further agree that any other questions, conflicts or matters of difference arising between them involving the rights, obligations, or interests of either in relation to the other or to the inhabitants of the other, across or along the boundary between the Parties, shall be referred to the Commission for evaluation and report, whenever any of the Parties shall request that such questions or matters of difference be so referred.

(c) (Optional, for use with §3A-1-03, Alternative 2) The Commissioners shall make a joint report to the respective Governments in all cases in which all or a majority of the Commissioners agree, and in case of disagreement the minority may make a joint report to both Governments, or separate reports to their respective Governments. Such reports of the Commission shall not be regarded as decisions of the questions or matters so submitted either on the facts or the law, and shall in no way have the character of an arbitral award. In case the Commission is evenly divided upon any question or matter referred to it for report, separate reports shall be made by the Commissioners on each side to their own Government.

(d) (Optional, for use with §3A-1-03, Alternative 2) Unanimity of the Commissioners shall be required to render a decision. In case the Commission is divided upon any question or matter presented to it for decision, the Commissioners on each side shall make separate reports to their own Government. The Parties shall thereupon endeavor to agree upon an adjustment of the question or matter of difference, and if an agreement is reached between them, it shall be reduced to writing in the form of a protocol, and shall be communicated to the Commissioners, who shall take such further proceedings as may be necessary to carry out such agreement.

Commentary: The Commission is charged with certain purposes that are specific to the Parties agreement to "the equitable and reasonable use of the water resources shared between the Parties; the exchange of data and other information pertinent to water utilization by the respective Parties; the cooperation and consultation necessary to achieve the purposes of this

Agreement." This provision provides the necessary special powers to achieve the goals of the Agreement.

Cross-references: §1A-1-04 (preservation of federal rights); §1A-1-05 (national security); §2A-1-02 (consent to jurisdiction); §2A-1-03 (duration of agreement); §2A-1-04 (amendments and supplements); §2A-1-05 (powers of sovereign parties; withdrawal); §2A-1-06 (existing agencies); §2A-1-07 (limited applicability); §2A-2-05 (party or parties); §3A-1-07 alternative 2, rules of procedures); §3A-2-01 (alternative 2, general powers and duties); §3A-3-01 (alternative 2, meetings, hearings and records§3A-3-02 (alternative 2, funding and expenses of the commission); §4A-1-01 (exclusive jurisdiction and control); §4A-1-02 (data exchange); Article 5A (dispute resolution).

Similar Agreements: *Treaty between the United States and Great Britain relating to Boundary Waters,* 36 Stat. 2451 (1909); Convention Between the Federal Republic of Germany and the Czech and Slovak Federal Republic and the European Economic Community on the International Commission for the Protection of the Elbe, International Environmental Law, Multilateral Agreements, 976:90/1 (1990).

Part 3 Administrative Procedures

§3A-3-01   MEETINGS, HEARINGS AND RECORDS (Optional)

(a) The signatory Parties recognize the importance and necessity of public participation in promoting utilization of the water resource of the _____ River Basin. Consequently, all meetings of the Commission shall be open to the public except with respect to issues of personnel.

(b) The minutes of the Commission shall be a public record open to inspection at the respective offices of the Commissioners or their alternates during regular business hours.

Commentary.   Although optional, inclusion of this provision is essential to maximum effectiveness of coordination and cooperation. Incorporation of this section memorializes the public nature of the enterprise. Effective water sharing demands that all stakeholders have information upon which they can rely in order for them to make rational decisions about water use. Without sufficient public participation, the Parties will be unable to maximize the use of the water resource. The US-Canada International Joint Commission has recognized the need for "engaging public support..." *See* International Joint Commission, Second Biennial Report under the Great Lakes Water Quality Agreement of 1978 to the Governments of the United States and Canada and the Provinces of the Great Lakes Basin (1984). Agenda 21 recognizes the need for the widest cooperation between governmental and non-governmental organizations.

Cross-references: §3A-1-07 (rules of procedures); §3A-2-01 (general powers and duties); §3A-2-02 (alternative 2, special powers and duties).

Similar Agreements: *Delaware River Basin Compact*, Pub. L. 87-328, 75 Stat. 688 (1961); *Susquehanna River Basin Compact*, Pub. L. No. 91-575, 84 Stat. 1509 (1970); *Apalachicola-Chattahoochee-Flint River Basin Compact*, O.C.G.A. 12-10-100 (1997); *Alabama-Coosa-Tallapoosa River Basin Compact*, O.C.G.A. 12-10-110 (1997); *Convention on the Protection and Use of Transboundary Watercourses and International Lakes*, 31 I.L.M. 1312 (1992); *The North American Agreement on Environmental Cooperation between the Government of the United States of America, the Government of Canada, and the Government of the United Mexican States*, 32 I.L.M. 1480 (1993).

§3A-3-02 FUNDING AND EXPENSES OF THE COMMISSION

Commissioners shall serve without compensation from the Commission. All general operational funding required by the Commission and agreed to by the members shall obligate each Party to pay an equal (equitable) share of such agreed upon funding. Funds remitted to the Commission by a Party in payment of such obligation shall not lapse; provided, however, that if any Party fails to remit payment within 90 days after payment is due, such obligation on the part of the other Parties shall terminate and any Party which has made payment may have such payment returned.

Commentary: Effective operation of the Commission, and hence effective implementation of the Agreement, requires a dependable source of funding. If a program of financing is uncertain, the Agreement will fail. Although this provision suggests equal funding by the Parties, other possibilities exist. "Equitable" funding may balance national economic power and the ability to pay for the Commission. Alternatively, "equitable" funding may base funding on some economic measure of the benefits accruing to each of the Parties as a result of the Agreement. The term once decided upon should be defined in Article 2, Part 2.

Cross-references: §3A-1-01 (alternative 2, commission created, alternative 2); §3A-1-02 (alternative 2, commission jurisdiction, alternative 2); §3A-1-03 (alternative 2, commissioners, alternative 2); §3A-1-04 (alternative 2, status, immunities and privileges, alternative 2); §3A-1-05 (alternative 2, commission organization and staffing); §3A-1-06 (alternative 2, advisory boards); §3A-3-01 (alternative 2, meetings, hearings and records).

Similar Agreements: Convention Between the Federal Republic of Germany and the Czech and Slovak Federal Republic and the European Economic Community on the International Commission for the Protection of the Elbe, International Environmental Law, Multilateral Agreements, 976:90/1 (1990); *The North American Agreement on Environmental Cooperation between the Government of the United States of America, the Government of Canada, and the Government of the United Mexican States*, 32 I.L.M. 1480 (1993); *Agreement on Cooperation for the Sustainable Development of the Mekong River Basin*, 34 ILM 864 (1995); *Apalachicola-Chattahoochee-Flint River Basin Compact*, O.C.G.A. 12-10-100 (1997); *Alabama-Coosa-Tallapoosa River Basin Compact*, O.C.G.A. 12-10-110 (1997).

ARTICLE 4A

COORDINATION OF WATER ISSUES

§4A-1-01 EXCLUSIVE JURISDICTION AND CONTROL

(a) Each of the Parties reserves to itself, unless otherwise mandated by U.S. federal law or contractually agreed upon by the Parties, the exclusive control over the utilization, consumption or diversion of all waters within its jurisdiction.

(b) The Parties covenant, however, that any interference with or diversion from its natural channel of shared waters which result in injury to the other Party's utilization of the shared waters shall give rise to the injured parties the same rights and entitlements to the same legal remedies as if such injury took place in the jurisdiction of the Party where such use, diversion or interference occurs.

(c) It is understood, however, that none of the Parties intends by the foregoing provision to surrender any right that it may have to enjoin or otherwise object to any interference with or diversion by the other Party of shared waters that has a reasonable potential to cause material injury to the utilization of shared waters within its jurisdiction.

Commentary: This provision establishes the principle of the sovereign right of each Party to allocate or otherwise utilize and control the waters within its jurisdiction, constrained only by the requirement that the such use be reasonable and equitable. Enforcement of this principle shall be based on those causes of actions and remedies available in tort within the legal system of the Party causing the injury. This section does provide for prospective relief from use, interference or diversion that may have a reasonable potential to cause material harm. A "reasonable potential" to cause material injury would be determined as a matter of law.

This provision provides legal remedies for the citizens of one Party who may be injured by the activities involving the shared waters by citizens of another Party. This provision provides the injured citizen with standing to sue, or standing to seek administrative remedies in the jurisdiction of another Party. Without such a provision, an injured individual would only be able to seek remedies by petitioning his/her own government to seek remedies from the other government. The complexities of such a political and administrative solution dramatically limit timely redress of the injury

Cross-references: §1A-1-04 (preservation of federal rights); §1A-1-05 (national security); §2A-1-02 (consent to jurisdiction); §2A-1-05 (powers of sovereign parties; withdrawal); §2A-2-05 (party or parties); §4A-1-02 (data exchange).

Similar Agreements: *Treaty between the United States and Great Britain relating to Boundary Waters, and Questions arising between the United States and Canada*, 36 Stat. 2451 (1909); *Sabine River Compact*, 68 Stat. 690 (1953).

§4A-1-02   DATA EXCHANGE

(a) The Paries agree to exchange complete documentation, data, and information on the status and condition of water in the various phases of the hydrologic cycle and on the development and utilization of those waters within their respective borders, if such development and utilization affect the shared water resources. The documentation, data and information shall be sufficient in scope to allow the other Parties to determine the potential damages that may therefrom arise. Such exchange shall include, but is not limited to, the following activities that may substantially raise or lower the historical levels, or change the natural periodic fluctuations, of the shared resource:

(1) Planned remedial or protective works or any dams, reservoirs or other diversions or obstructions to waters;

(2) Significant interbasin transfers;

(3) Changes to consumptive water utilization;

(4) Existing and planned flood protection programs and works which increase the risk of flood damages to other Parties;

(5) Water development or utilization that has an adverse effect on the biological, physical and chemical quality of the shared resource, so as to affect the public health and safety, the recreational potential, environmental sustainability or the quality of life of the other Parties.

(6) Any augmentation of water supply that may affect the other party.

(b) (Optional) Meetings to exchange complete documentation, data and information shall take place, upon the call of the Party originating such data, within 60 days of such data being avaiable. Exchange of data and any consultation resulting from provision of the data shall occur prior to the data and information becoming part of the public record.

Commentary: Water utilization by one Party can have serious effects on the quantity and/or quality of the water availble for use by the other Parties. A significant source of controversy develops when one Part undertakes The construction and operation of water supply reservoirs and hydropower facilities may may dramatically lower the flow in the shared water resource. Although this reduction may be limited to the initial start-up period and may be limited to a period of several years, severe economic and social impact may occur to other Parties. As importantly, releases from the works must be coordinated to assure downstream users are not affected. In the case of water supply reservoirs this may become critical during periods of drought. In the case of hydropower dams, especially those of a "peaking power" nature, the timing of release may be critical. Other Parties may also be affected by significant changes resulting from interbasin transfers or increased water consumption.

Flood control policies and works also have a dramatic effect on the timing and elevation of water levels and thus may become a major contentious issue between the Parties. The issue should be addressed as an individual area of coordination. This provision recognizes the sovereign right of each Party to make efforts to safeguard its people and economic forces from flood damages but establishes an avenue for the sharing of data on flood control efforts as well as an independent analysis of the effects of those efforts on other Parties.

Changes and/or additions to the industrial base of one Party, or increased urbanization of a Party, can result in substantially degraded discharges into the shared waters. Poor quality water imposes risks that the Parties should recognize as a common threat. First there is the health risk to the population that uses the water for domestic purposes. Second if the available water will not meet the standards for certain industrial purposes, there is the risk that economic growth will be impaired. Finally there is the risk that quality degradation will have a severe impact on the ecology of the Basin, resulting in long-term sustainability complications. Integration of water quality and quantity is essential. Agenda 21 obligated all signatories to develop a program of water and sustainable development; see also A. Satre Ahlander, Environmental Policies in the former Soviet Union, Economic Policies for Sustainable Development, (Thomas Sterner, ed., 1994) The UN *Convention on the Law of the Non-Navigational Uses of International Watercourses*, United Nations Document A/51/869 (1998), establishes the criterion that "(w)atercourse States shall, individually and, where appropriate, jointly, protect and preserve the ecosystems of international watercourses."

Surface and underground water may be degraded by a variety of factors. Major problems affecting the quality of these water resources arise, for instance, from inadequate domestic sewage treatment, inadequate controls on the discharge of industrial waste and effluent, the diversion of waters resulting in insufficient water to assimilate waste, the loss and destruction of catchment areas, the improper siting of industrial plants, deforestation, and poor agricultural practices that cause leaching of nutrients and pesticides. Transboundary water sharing must include effective plans and programs that eliminate, or at least minimize, the possible sources of water quality degradation.

The timing of the data exchange is important for all Parties. It is recommended that the data and information be provided to other Parties prior to the information becoming a part of the public record since the capability for changing public plans and programs are difficult at best.

Cross-references: §1A-1-04 (preservation of federal rights); §1A-1-05 (national security); §2A-1-02 (consent to jurisdiction); §2A-2-03 (flood); §2A-2-04 (interbasin transfer); §2A-2-05 (party or parties); §4A-1-01 (exclusive jurisdiction and control);

Similar Agreements: *Klamath River Basin Compact*, 71 Stat. 497 (1957); Arkansas River Basin Compact of 1965, 80 Stat. 1409 (1966); Big Blue River Basin Compact, 86 Stat. 193 (1972); Arkansas River Basin Compact of 1970, 87 Stat. 569 (1973); *Delaware River Basin Compact*, Pub. L. 87-328, 75 Stat. 688 (1961); *Susquehanna River Basin Compact*, Pub. L. No. 91-575, 84 Stat. 1509 (1970); *Kansas-Nebraska Big Blue River Compact*, 86 Stat. 193 (1972); *Red River Compact*, 94 Stat. 3305 (1980); Oregon-California Goose Lake Interstate Compact, 98 Stat. 291 (1984); *Treaty between the United States and Great Britain relating to Boundary Waters, and Questions arising between the United States and Canada*, 36 Stat. 2451 (1909); *Treaty between the United States of America and Mexico, Utilization of Waters of the Colorado and Tijuana rivers and of the Rio Grande*, 59 Stat. 1219 (1945); *Pecos River Compact* (1948)Agreement between the United States and Canada on Great Lakes Water Quality, 1153 UNTS 187 (1978); *Convention on the Protection and Use of Transboundary Watercourses and International Lakes*, 31 I.L.M. 1312 (1992); *Convention on the Law of the Non-Navigational Uses of International Watercourses*, United Nations Document A/51/869 (1998).

ARTICLE 5A

DISPUTE RESOLUTION

(OPTIONAL)

Commentary. Disputes will inevitably arise as the Agreement is implemented. Thesed disputes may involve differences in interpretation of the Agreement's provisions or non-compliance with the Agreement itself. The disputes may also arise because of changing conditions that alter the effectiveness of the Agreement for one or more of the Parties. While a speedy and equitable process of dispute resolution serves all Parties well, some sovereign entities do not wish to enter into an obligatory process. In such a case, Article 5A may be omitted. In other instances, the Parties may recognize the need to institutionalize a dispute resolution process

§5A-1-01 RESOLUTION BY SIGNATORY PARTIES

Whenever any difference or dispute may arise between two or more Parties to this Agreement regarding any matters covered by this Agreement, particularly as to the interpretations of the Agreement and the legal rights of the parties, the Parties shall first make every effort to resolve the issue through negotiations and consultations based on the powers and duties herein described.

Commentary. This alternative is the least restrictive upon the sovereignty of the Parties yet expresses their recognition of the need for peaceful resolution of dipustes. The alternative is appropriate in those cases where no commission has been established to manage the data exchange.

Cross-references: §1A-1-01 (general policies); §1A-1-02 (coordination and cooperation); §1A-1-03 (good faith implementation); §1A-1-04 (preservation of federal rights); §1A-1-05 (national security); §2A-1-05 (powers of sovereign parties; withdrawal); §2A-1-06 (existing agencies); §2A-1-07 (limited applicability); §2A-2-05 (party or parties); §5A-1-02 (alternative 1, right to litigate).

Similar Agreements: *Agreement on Cooperation for the Sustainable Development of the Mekong River Basin*, 34 ILM 864 (1995).

§5A-1-02   RIGHT TO LITIGATE

Nothing in this agreement shall be construed to limit or prevent either Party from instituting or maintaining any action or proceeding, legal or equitable, in any tribunal of competent jurisdiction for the protection of any right under this agreement or the enforcement of any of its provisions.

Commentary: The existence of an appropriate tribunal may pose a problem in cases not involving an entity like the United States or European Union. It may be advisable to specify the tribunal in the agreement itself to avoid dispute over jurisdictional questions at a later date.

Cross References: §1A-1-01 (general policies); §1A-1-02 (coordination and cooperation); §1A-1-03 (good faith implementation); §1A-1-04 (preservation of federal rights); §1A-1-05 (national security); §2A-1-05 (powers of sovereign parties; withdrawal); §2A-1-06 (existing agencies); §2A-1-07 (limited applicability); §2A-2-05 (party or parties); §5A-1-01 (alternative 1, resolution by signatory parties,).

Similar Agreements: *Belle Fourche River Compact*, 1944; *Colorado River Compact*, 1928; *Snake River Compact*, 1949.

SIGNATURES

IN WITNESS WHEREOF, and in evidence of the adoption and enactment into law of this Agreement by the signatory Parties, the representatives of the sovereign States of _____. _____, and _____ do hereby, in accordance with authority conferred by law, sign this Agreement in (____) duplicate original copies, as attested by the appropriate authorities of the respective sovereign Parties, and have caused the seals of the respective States to be hereunto affixed this ____ day of _____.

LIMITED PURPOSE AGREEMENT CONCERNING THE SHARED USE OF TRANSBOUNDARY WATER RESOURCES

Model B

Contents

Limited Purpose Agreement Concerning the Shared Use of Transboundary Water Resources

Introduction...41
Article IB – Declaration of Policies and Purposes 42
Purposes and Scope of Agreement..42
Coordination and Cooperation..43
Good Faith Implementation..45
Preservation of Federal Rights (Optional)..46
National Security (Optional, for International Use)...47

Article 2B – General Provisions 47
Part 1: General Obligaitons..47
Effective Date...47
Duration Agreement (Optional)..47
Consent of Jurisdiction (for U.S. Use Only)...49
Amendments and Supplements (Optional)...50
Limited Applicability (Optional)..50
Annexes 51
Part 2: Definitions...52
Atmospheric Water 52
_____ Basin 52
Conservation Measures...52
Drought 53
Equitable and Reasonable Utilizations, Consumption or Diversion........................53
Flood 54
Party or Parties..55
Underground Water...55
Waters of the Basin 55

Article 3B – Administration 56
Part 1: Administration Officials 56
Use of Water Management Officials of the Parties................................... 56
Substitution of Officials... 57
Implementation and Verification of Agreement...................................... 57
Funding.. 58
Part 2: Activities Within the Territory of Other Parties 58
Rights in Territory of Other Party (Optional)... 58
Storage and Diversion (Optional)... 59
Eminent Domain (Optional)... 59
Navigation Servitude (for International Use) 61

Article 4B – Equitable and Reasonable Use of Water 62
Exclusive Jurisdiction and Control... 62
Water Allocation (Optional).. 63
Water Levels Protected (Optional).. 64
Underground Water; Limit on Withdrawals (Optional).......................... 65
Flood Protection Works (Optional).. 65
Augmentation of Supply (Optional)... 66
Water Quality (Optional)... 67

Article 5B – Dispute Resolution 69
Resolution by Signatory Parties.. 70
Rights to Litigate 70

Signatures 71

ANNEX B-1: Allocation Alternative 72

INTRODUCTION

Model B, the Limited Purpose Agreement for the Shared Use of Transboundary Water Resources document, is designed for those situations in which the Parties wish to maintain control of most aspects of their internal water development but recognize either the need to resolve existing or potential conflict or the need to establish direct coordination or management over a specific water development project a particular water source, or a particular water management function. Essentially, this model is oriented towards purposes that are limited in scope and narrowly drawn. The limited purpose goals may vary from simple allocation of water from a particular water source to other matters such as pollution control of a specific water body. Model B establishes an agreement for the joint development of specific water resources projects, the management of which may require more than simple coordination and cooperation. However, the joint development is strictly limited to the specific development project rather than being a comprehensive program that is expansive in nature. Compare Model A (Coordination and Cooperation) and Model C (Comprehensive Management). This model agreement therefore presents a number of illustrative provisions, not all of which may be appropriate for use in a given case. The provisions set out below can serve as guidelines based on similar provisions found in existing water-sharing agreements in the United States.

This model agreement is intended for use when a specific product is sought in terms of water allocation or water quality control. The operative provisions that actually apportion the water resources or define water quality standards will of necessity vary with the particular circumstances. The following provisions are based upon existing agreements, with commentary explaining advantages and disadvantages, and may be used as models, in whole, in combination, or in part, for new agreements.

LIMITED PURPOSE AGREEMENT
CONCERNING THE
SHARED USE OF TRANSBOUNDARY WATER RESOURCES
OF THE
_____ RIVER BASIN

ARTICLE 1B

DECLARATION OF POLICIES AND PURPOSES

§1B-1-01 PURPOSES AND SCOPE OF AGREEMENT

(a) The waters of the _____ River Basin have local, regional and national significance and equitable and reasonable allocation of the shared waters of the _____ River are public purposes for the respective signatory Parties.

(b) The purposes of this *A*greement are to promote interstate [international] comity; to remove causes of present and future controversy; to make secure and protect water resouce developments within the Parties; equitably and reasonably allocate the shared waters of the _____ River; and to augment the benefits of the shared waters of the _____ River Basin through joint planning and management of specific projects.

(c) The physical and other conditions peculiar to the _____ River Basin constitute the basis for this Agreement, and its provisions are applicable only [to the surface waters of the Basin] [to underground waters and atmospheric water augmentation as well as the surface waters affecting the Basin].

Commentary: This section is critical to future interpretation and implementation of the *A*greement. The nature of this product-oriented compact presupposes the existence of specific purposes for which the agreement is made, and those should be set out here. If the agreement is to focus only on allocation, for example, references to drought or flood control strategies would be excluded.

It is important to acknowledge that the Agreement reflects the particular circumstances and compromises reached in its formulation, as applied to the particular basin. Care should be taken to ensure that this Agreement cannot be applied to other Basins, unless the intent of the Parties is otherwise. The inclusion of §1-1-01(c) avoids later claims that other rivers and basins, or other bodies of water, should be dealt with in a similar manner. If underground water and atmospheric water are to be included within the scope of the agreement, it may be mentioned here. It is particularly important to address the atmospheric and underground water issues in this paragraph to avoid later disputes over whether or not underground water and atmospheric water are included within the scope of the Agreement.

Cross References: §1B-1-02 (coordination and cooperation); §1B-1-03 (good faith implementation); §1B-1-04 (preservation of federal rights); §1B-1-05 (national security); §2B-1-01 (effective date); §2B-1-02 (duration of agreement); §2B-1-03 (consent to jurisdiction); §2B-1-04 (amendments and supplements); §2B-1-05 (limited applicability); §2B-1-06 (annexes); §2B-2-01 (atmospheric water); §2B-2-03 (conservation measures); §2B-2-04 (drought); §2B-2-06 (flood); §2B-2-07 (party or parties); §2B-2-08 (underground water); §2B-2-09 (waters of the basin); §3B-1-01 (use of party officials); §3B-1-02 (substitution of officials); §3B-1-03 (implementation and verification of agreement); §3B-1-04 (funding); §3B-2-01 (rights in territory of other party); §3B-2-02 (storage and diversion); §3B-2-03 (eminent domain); §3B-2-04 (navigational servitude); §4B-1-01 (exclusive jurisdiction and control); §4B-1-02 (water allocation); §4B-1-03 (water levels protected); §4B-1-04 (underground water; limit on withdrawals); §4B-1-05 (flood protection works); §4B-1-06 (augmentation of supply); §4B-1-07 (water quality).

Similar Agreements: *Arkansas River Compact*, 63 Stat. 145 (1949); *Belle Fourche River Compact*, 58 Stat. 94 (1944); *Canadian River Compact*, 66 Stat. 74 (1952); *Colorado River Compact*, approved 45 Stat. 1057 (1928); text found in Congressional Record, 10 December 1928, 32`325, Art. I; *Costilla Creek Compact*, 60 Stat. 246 (1946); amended 77 Stat. 350, Art. I (1963); *Kansas- Nebraska Big Blue River Compact*, 86 Stat. 193 (1972); *Klamath River Basin Compact*, 71 Stat. 497 (1957); *Pecos River Compact*, 63 Stat. 159 (1948); *Red River Compact*, 94 Stat. 3305 (1978); *Snake River Compact*, 64 Stat. 29 (1949); *Upper Colorado River Basin Compact*, 63 Stat. 31 (1948); *Upper Niobrara River Compact*, 83 Stat. 86 (1969).

§1B-1-02 COORDINATION AND COOPERATION

Alternative 1

(a) Each of the Parties pledges to support implementation of the provisions of this Agreement, and covenants that its officers and agencies will not hinder, impair, or prevent any other Party carrying out any provision or recommendation of this Agreement.

(b) The Parties shall at all times endeavor to agree on the interpretation and application of this Agreement, and shall make every attempt through cooperation

and consultations to arrive at a mutually satisfactory resolution of any matter that might affect its operation.

(c) The Parties agree that their respective governmental organizations shall provide the information necessary to assist in the equitable and reasonable utilization of those resources. Such information shall include, but not be limited to, all planning and management activities and water projects affecting their shared water resources.

(d) The Parties further acknowledge that all states are expected to conduct themselves with an absence of malice and deceit, with no intention to seek unconscionable advantage.

Alternative 2

The parties agree to the following objectives:

(a) To cooperate and consult with the other Parties to this Agreement in their development, utilization, consumption and conservation of the water and related resources shared by the Parties in order to ensure equitable and reasonable use of those waters while minimizing harm to other Parties.

(b) To cooperate on the basis of sovereign equality and territorial integrity in the utilization and protection of the shared water resources.

(c) The Parties further acknowledge that all states are expected to conduct themselves with an absence of malice or deceit and with no intention to seek unconscionable advantage.

Commentary: Normally a state or nation enters into any international agreement with a position of self-interest. In the negotiations, each Party seeks the rights and authorities critical to certain political, economic or social objectives while ceding less critical rights and authorities to the other Parties. While accepting this fact, all Parties have a duty to cooperate and negotiate in good faith. This principle is the foundation of international law, and it applies in all relations between sovereign states.

This provision provides a framework for the Parties in their development of their individual water policy planning. It recognizes that there are certain fundamental principles that each Party must follow in their rationale management of water resources. It would be irrational for one Party to agree to "equitable and reasonable utilization" when it does not follow a similar philosophy within its own borders. These general objectives and principles improve the likelihood of accomplishing of purposes of the water sharing.

Cross-references: §1B-1-01 (purposes and scope of agreement); §1B-1-03 (good faith implementation); §1B-1-04 (preservation of federal rights); §1B-1-05 (national security); §2B-1-01 (effective date); §2B-1-02 (duration of agreement); §2B-1-03 (consent to jurisdiction); §2B-1-

04 (amendments and supplements); §2B-1-05 (limited applicability); §2B-1-06 (annexes); §2B-2-05 (equitable and reasonable utilization); §3B-1-01 (use of party officials); §3B-1-02 (substitution of officials); §3B-1-03 (implementation and verification of agreement); §3B-1-04 (funding); §3B-2-01 Rights in Territory of Other Party; §3B-2-02 (storage and diversion); §3B-2-03 (eminent domain); §3B-2-04 (navigational servitude); §4B-1-01 (exclusive jurisdiction and control); §4B-1-02 (water allocation); §4B-1-03 (water levels protected); §4B-1-04 (underground water; limit on withdrawals); §4B-1-05 (flood protection works); §4B-1-06 (augmentation of supply); §4B-1-07 (water quality); §5B-1-01 (resolution by signatory parties); §5B-1-02 (right to litigate).

Similar Agreements: *Agreement Between the People's Republic of Bulgaria and the Republic of Turkey Concerning Co-operation in the Use of the Waters of Rivers Flowing through the Territory of Both Countries*, UNTS, Vol. 807, 117 (1968); *Convention Between Switzerland and Italy Concerning the Protection of Italo-Swiss Waters Against Pollution*, UNTS, Vol. 957, 277 (1972); *Stockholm Declaration of the United Nations Conference on the Human Environment*, 11 I.L.M. 1416 (1972); *Treaty for Amazonian Cooperation*, 17 ILM 1046 (1978); *Convention Between the Federal Republic of Germany and the Czech and Slovak Federal Republic and the European Economic Community on the International Commission for the Protection of the Elbe, International Environmental Law, Multilateral Agreements*, 1976:90/1 (1990); *Convention on the Protection and Use of Transboundary Watercourses and International Lakes*, 31 I.L.M. 1312 (1992); *The North American Agreement on Environmental Cooperation between the Government of the United States of America, the Government of Canada, and the Government of the United Mexican States*, 32 I.L.M. 1480 (1993); *Treaty of Peace between the State of Israel and the Hashemite Kingdom of Jordan*, 34 I.L.M. 43 (1994); *Agreement on the Cooperation for the Sustainable Development of the Mekong River Basin*, 34 ILM 864 (1995); *Convention on the Law of the Non-Navigational Uses of International Watercourses*, United Nations Document A/51/869 (1998).

§1B-1-03 GOOD FAITH IMPLEMENTATION

The Parties agree to implement all provisions of this Agreement, and each covenants that its officers and agencies will not hinder, impair, or prevent any other Party carrying out any provision or recommendation of this Agreement.

Commentary. It should be noted that good-faith misinterpretation of compact obligations do not excuse a Party from damage liability. *See Texas v. New Mexico*, 482 U.S. 124 (1987). In that case, the Supreme Court reasoned that a compact is a contract, and standard contract law does not allow a defense based on misinterpretation of contract obligations. *See* Grant, §45.07(c), §46.05(d).

Cross-references: §1B-1-01 (purposes and scope of agreement); §1B-1-02 (coordination and cooperation); §1B-1-04 (preservation of federal rights); §1B-1-05 (national security); §2B-1-01 (effective date); §2B-1-02 (duration of agreement); §2B-1-03 (consent to jurisdiction); §2B-1-04 (amendments and supplements); §2B-1-05 (limited applicability); §2B-1-06 (annexes); §3B-1-01 (use of party officials); §3B-1-02 (substitution of officials); §3B-1-03 (implementation and verification of agreement); §3B-1-04 (funding); §3B-2-01 (rights in territory of other party); §3B-2-02 (storage and diversion); §3B-2-03 (eminent domain); §3B-2-04 (navigational servitude); §4B-1-01 (exclusive jurisdiction and control); §4B-1-02 (water allocation); §4B-1-03 (water levels protected); §4B-1-04 (underground water; limit on withdrawals); §4B-1-05 (flood protection works); §4B-1-06 (augmentation of supply); §4B-1-07 (water quality); §5B-1-01 (resolution by signatory parties); §5B-1-02 (right to litigate).

Similar Agreements: *Helsinki Rules on the Uses of the Waters of International Rivers*, 52 I.L.A. 484 (1966); *Stockholm Declaration of the United Nations Conference on the Human Environment*, 11 I.L.M. 1416 (1972); *Convention on the Law of the Non-Navigational Uses of International Watercourses*, United Nations Document A/51/869 (1998).

§1B-1-04 PRESERVATION OF FEDERAL RIGHTS (Optional)

Nothing in this agreement shall be deemed:

(a) To impair or affect any rights or powers of the United States, its agencies or instrumentalities, in and to the use of the waters of the _____ River Basin nor its capacity to acquire rights in and to the use of said waters;

(b) To subject any property of the United States, its agencies, or instrumentalities to taxation by either State or subdivision thereof, nor to create an obligation on the part of the United States, its agencies, or instrumentalities, by reason of the acquisition, construction or operation of any property or works of whatsoever kind, to make any payments to any State or political subdivision thereof, State agency municipality, or entity whatsoever in reimbursement for the loss of taxes;

(c) To subject any property of the United States, its agencies, or instrumentalities, to the laws of any State to an extent other than the extent to which these laws would apply without regard to the *A*greement.

Commentary: This section should be included in agreements between states of the United States. Arguably, it may be unnecessary to preserve Federal rights, but inasmuch as Congress must approve the agreement, the inclusion of these provisions may facilitate obtaining that approval.

Cross References: §1B-1-01 (purposes and scope of agreement); §1B-1-05 (national security); §4B-1-01 (exclusive jurisdiction and control).

Similar Agreements: *Republican River Compact*, 57 Stat. 86 (1943); *Belle Fourche River Compact*, 58 Stat. 94 (1944); *Pecos River Compact*, 63 Stat. 159 (1948); *Snake River Compact*, 64 Stat. 29 (1949); *Upper Colorado River Basin Compact*, 63 Stat. 31 (1949); *Yellowstone River Compact*, 65 Stat. 663 (1950); *Canadian River Compact*, 66 Stat. 74, (1952); *Klamath River*

Basin Compact, 71 Stat. 497 (1957); *Bear River Compact*, 72 Stat. 38 (1955), amended 94 Stat. 4, Art. XIII (2) (1980).

§1B-1-05 NATIONAL SECURITY (Optional, for international use)

(a) Nothing in this Agreement shall be construed to require any Party to make available or provide access to information the disclosure of which it determines to be contrary to its essential security interests.

(b) Nothing in this Agreement shall be construed to prevent any Party from taking any actions that it consider necessary for the protection of its essential security interests relating to a formal declaration of war.

Commentary: National security concerns will necessarily take precedence over any program of water management and the exchange of data.

Cross-references: §1B-1-01 (purposes and scope of agreement); §1B-1-04 (preservation of federal rights); §4B-1-01 (exclusive jurisdiction and control).

Similar Agreements: *The North American Agreement on Environmental Cooperation between the Government of the United States of America, the Government of Canada, and the Government of the United Mexican States*, 32 I.L.M. 1480 (1993).

ARTICLE 2B

GENERAL PROVISIONS

Part 1 General Obligations

§2B-1-01 EFFECTIVE DATE

Alternative 1: (For International use)

This Agreement shall become operative when approved by the appropriate governing authorities of all Parties. The agreement will go into full force and effect at 12:01 a.m. [time zone] on the day immediately following the final act necessary for approval of the agreement, as defined by the domestic law of each Party, by the last Party to give such approval.

Alternative 2: (For U.S. use)

This agreement shall become operative when, sibsequent to approval by the Legislature of each of the States, the Congress of the United States adopts legislation providing, among other things, that:

    (a) Any equitable and reasonable uses hereafter made by the United States, or those acting by or under its authority, within a State, of the waters allocated by this agreement, shall be within the allocations hereinabove made for use in that State and shall be taken into account in determining the extent of use within that State.

    (b) The United States shall recognize, to the extent consistent with the best utilization of the waters for multiple purposes, that equitable and reasonable use of the waters within the Basin is of paramount importance to development of the Basin. This shall pertain to the exercise of rights or powers arising from whatever jurisdiction the United States has in, over and to the waters of the _____ River and all its tributaries. The United States government shall exercise no power that may interfere with the full equitable and reasonable use of the waters unless the exercise of such power is in the interest of the best utilization of such waters for multiple purposes.

Commentary: Any agreement of this nature should specify the date or conditions upon which it will take effect. In the case of agreements between states of the United States, the conditions with respect to Congress are designed to provide some measure of protection against subsequent federal action that might disturb the allocation system agreed upon by the contracting Parties. Despite the requirement of federal approval of interstate compacts, the federal government is not normally a Party to those agreements and may not be bound by the provisions of those agreements unless there is specific legislation committing the federal government to be so bound. The provisions of § 2-1-01, modeled after the *Republican River Compact*, 57 Stat. 86 (1943) and the *Bell Fourche Compact*, 58 Stat. 94 (1944) condition the effectiveness of the agreement on passage of such legislation by Congress and also establish a basis for compensation for takings under the Fifth Amendment should a subsequent Congress decide to take action contrary to that commitment. A later Congress has the power to set aside the actions of an earlier Congress, but the question of takings and just compensation then arises. If these conditions are not incorporated, the States making the agreement may later find that federal actions render their agreement ineffective.

Cross-references: §1B-1-01 (purposes and scope of agreement); §2B-1-02 (duration of agreement); §2B-1-03 (consent to jurisdiction); §2B-1-04 (amendments and supplements); §2B-1-05 (limited applicability); §2B-1-06 (annexes); §3B-1-01 (use of party officials); §3B-1-02 (substitution of officials); §3B-1-03 (implementation and verification of agreement).

Similar Agreements: *Republican River Compact*, 57 Stat. 86 (1943); and Bell Fourche *River Compact*, 58 Stat. 94 (1944); *Delaware River Basin Compact*, Pub. L. 87-328, 75 Stat. 688 (1961); *Susquehanna River Basin Compact*, Pub. L. No. 91-575, 84 Stat. 1509 (1970); *Convention on the Protection and Use of Transboundary Watercourses and International Lakes*, 31 I.L.M. 1312 (1992); *Convention on the Law of the Non-Navigational Uses of International Watercourses*, United Nations Document A/51/869 (1998).

§2B-1-02 DURATION AGREEMENT (Optional)

The Parties intend that the duration of this Agreement shall be for an initial period of [___] years from its effective date.

Commentary: The Parties may prefer to establish no duration and rely on later provisions to modify or terminate the agreement. However, two significant principles are established by this provision. First, setting a duration for an extended period of time, 25 0r 50 years, allows for predictability on terms of water resources development; it also allows sufficient time to recover capital costs in the financing of projects. Second, establishing an extended duration ensures that the Parties reconsider the Agreement only after a sufficient hydrologic record is established. However, an extended duration does somewhat constrain the Parties if significant climate change occurs and dramatically alters the hydrology of the shared water resource. Additionally, significant changes in water demands or changes in water technology could make the terms of the agreement unworkable.

Cross-references: §2B-1-01 (effective date); §2B-1-04 (amendments and supplements).

Similar Agreements: *Delaware River Basin Compact* (DRBC), Pub. L. 87-328, 75 Stat. 688 (1961); *Susquehanna River Basin Compact*, Pub. L. No. 91-575, 84 Stat. 1509 (1970).

§2B-1-03 CONSENT TO JURISDICTION (for U.S. use only)

This agreement shall be effective upon the United States Congress giving its consent for the United States to be named and joined as a Party defendant or otherwise in any case or controversy involving the construction or application of this agreement in which one or more of the States is a plaintiff, without regard to any requirement as to the sum or value in controversy or diversity of citizenship of Parties to the case or controversy.

Commentary: The predominance of federal interests in water resources makes it likely that any litigation concerning the agreement between States will involve federal interests. The doctrine of sovereign immunity could prevent joinder of the federal interests as Parties to the suit absent a waiver of sovereign immunity. The inability to join federal parties led to dismissal of a suit filed by Texas against New Mexico in 1951 to enforce certain provisions of the *Rio Grande Compact* of 1938, 53 Stat. 785, 938. The Supreme Court dismissed the case because the federal government was not joined as a party, but had important interests that would be affected by any such suit. *Texas v. New Mexico*, 352 U.S. 991, 957. The Parties should consider including such a waiver of sovereign immunity as a condition to effectiveness of the Agreement. They may also wish to add a provision granting jurisdiction over any such cases to the District Courts, which may be preferable to the Supreme Court as the initial forum for resolving certain types of disputes. *The Red River Compact*, 94 Stat. 3305 (1980) takes this approach.

Cross References: §1B-1-04 (preservation of federal right); §1B-1-05 (national security).

Similar Agreements: *Kansas-Nebraska Big Blue River Compact*, 86 Stat. 193 (1972); *Red River Compact*, 94 Stat. 3305 (1980).

§2B-1-04   AMENDMENTS AND SUPPLEMENTS (Optional)

The provisions of this *A*greement shall remain in full force and effect until amended by action of the governing bodies of the Parties and consented to and approved by any other necessary authority in the same manner as this agreement is required to be ratified to become effective.

Commentary.   Agreements may, over time, fail to operate as well as initially intended. Therefore, some amendment process should be specified. In some cases, the approval of another institution may be required. If, for example, the agreement is between states of the United States, the United States Constitution arguably requires Congressional approval of any amendment as well as approval of the original agreement, unless the agreement provides for a different method of amendment. In this latter case, The Congressional approval of the initial agreement would implicitly grant consent to modify the agreement in accordance with the terms of the agreement. If the agreement is between sovereign nations, the references to other "necessary authority" may be omitted, but the particular circumstances of each case must be considered.

Cross-references: §1B-1-01 (purposes and scope of agreement); §2B-1-05 (limited applicability); §2B-1-06 (annexes); §3B-1-03 (implementation and verification of agreement).

Similar Agreements: *Delaware River Basin Compact*, Pub. L. 87-328, 75 Stat. 688 (1961); *Susquehanna River Basin Compact*, Pub. L. No. 91-575, 84 Stat. 1509 (1970); *Convention on the Protection and Use of Transboundary Watercourses and International Lakes*, 31 I.L.M. 1312 (1992); *Agreement on the Cooperation for the Sustainable Development of the Mekong River Basin*, 34 ILM 864 (1995).

§2B-1-05   LIMITED APPLICABILITY (Optional)

Should a tribunal of competent jurisdiction hold any part of this agreement to be void or unenforceable, all other severable provisions shall continue in full force and effect.

Commentary: The drafters of the agreement should consider whether they wish this clause to be included. The advantage of such a clause is that it avoids the possibility of having the entire agreement become null and void if any part is found to be void or unenforceable. On the other hand, the agreement may be viewed as such an integrated package that the Parties would choose to have the entire agreement fall if any part falls.

Cross References: §1B-1-01 (purposes and scope of agreement); §2B-1-04 (amendments and supplements); §2B-1-06 (annexes); §3B-1-01 (use of party officials); §3B-1-02 (substitution of officials); §3B-1-03 (implementation and verification of agreement); §3B-1-04 (funding); §3B-2-01 (rights in territory of other party); §3B-2-02 (storage and diversion); §3B-2-03 (eminent domain); §3B-2-04 (navigational servitude); §4B-1-01 (exclusive jurisdiction and control); §4B-1-02 (water allocation); §4B-1-03 (water levels protected); §4B-1-04 (underground water; limit on withdrawals); §4B-1-06 (augmentation of supply); §4B-1-07 (water quality); §5B-1-01 (resolution by signatory parties); §5B-1-02 (right to litigate).

Similar Agreements: *Yellowstone River Compact,* 65 Stat. 663 (1950); *Sabine River Compact,* 68 Stat. 690 (1953); *Klamath River Basin Compact,* 71 Stat. 497 (1957); *Delaware River Basin Compact,* Pub. L. 87-328, 75 Stat. 688 (1961); *Susquehanna River Basin Compact,* Pub. L. No. 91-575, 84 Stat. 1509 (1970).

§2B-1-06 ANNEXES

The Annexes to this Agreement, to the extent that they are consistent with the objectives and intent of the Agreement, constitute an integral part of the Agreement.

Commentary: An effective water management agreement will necessarily contain detailed information and data of a procedural nature. While such information may be essential to the effectiveness of the particular agreement, its inclusion in the main body of the agreement may take away from the essence of the contractual nature of the agreement. The use of annexes minimizes this effect.

Cross-references: §2B-1-04 (amendments and supplements); §3B-1-01 (use of party officials); §3B-1-02 (substitution of officials); §3B-1-03 (implementation and verification of agreement); §3B-1-04 (funding); §3B-2-01 (rights in territory of other party); §3B-2-02 (storage and diversion); §3B-2-03 (eminent domain); §3B-2-04 (navigational servitude); §4B-1-01 (exclusive jurisdiction and control); §4B-1-02 (water allocation); §4B-1-03 (water levels protected); §4B-1-04 (underground water; limit on withdrawals); §4B-1-06 (augmentation of supply); §4B-1-07 (water quality); §5B-1-01 (resolution by signatory parties); §5B-1-02 (right to litigate).

Similar Agreements: ASEAN Agreement on the Conservation of Nature and Natural Resources, 1985; *The North American Agreement on Environmental Cooperation between the Government of the United States of America, the Government of Canada, and the Government of the United Mexican States,* 32 I.L.M. 1480 (1993).

Part 2 Definitions

§2B-2-01 ATMOSPHERIC WATER

The phrases "atmospheric water" means water produced by artificial changes in the composition, motions, and resulting behavior of the atmosphere or clouds, including fog, or with the intent of inducing changes in precipitation by use of electrical device, lasers, alterations of the earth's surface, or cloud seeding.

Commentary: This definition is consistent with the definitions usually used in State and federal laws on weather modification. See generally Robert Beck, *Augmenting the Available Water Supply*, in 1 Waters and Water Rights § 3.04; Ray Jay Davis, *Four Decades of American Weather Modification Law,* 19 J. Weather Modification 102 (1987).

Cross-references: §1B-1-01 (purposes and scope of agreement); §2B-2-09 (waters of the basin).

§2B-2-02 _____ BASIN

"_____ Basin" means the area of drainage into the _____ River and its tributaries, [and] aquifers underlying the drainage, or only the aquifers themselves.

Commentary: The Agreement could include the total surface area of drainage throughout the Basin and contain aquifers underlying the surface drainage. Some tributaries can be connected to the underlying aquifers holding the underground water. Some of the aquifers could be connected to more than one of the surface water basins. The geographic scope of the Agreement should be defined to ensure there are no future disagreements about what lands are or are not covered by the Agreement. A map may be incorporated, but care should be taken that the map is cartographically accurate. Because the map is likely to be at a scale too small for precise delineation of boundaries, it should be made clear that it is for general reference only. In the event of a dispute over land or within the defined _____ River and its tributaries, the actual limits of the watershed as determined on the ground should be controlling.

Cross-references: §2B-2-05 (equitable and reasonable utilization); §3B-1-02 (Substitution of Officials); §4B-1-03 (Water Levels Protected).

§2B-2-03 CONSERVATION MEASURES

"Conservation measures" refers to any measures adopted by a water right holder, or several water right holders acting in concert pursuant to a conservation agreement reviewed and approved by the Commission as being appropriate water-saving strategies for purposes of the Comprehensive Water Management Plan, to reduce the withdrawals and/or consumptive uses, including, but not limited to:

(a) **Improvements in water transmission and water use efficiency;**
(b) **Reduction in water use;**
(c) **Enhancement of return flows; and**
(d) **Reuse of return flows.**

Commentary: Sustainable development requires steps to conserve the waters of the Stat basin. This definition limits the application of the term "conservation measures" to practices that have been reviewed and approved by the Commission as being appropriate water-saving strategies for the purposes of the Comprehensive Water Management Plan. Specifically excluded from this definition are practices applied to native or naturally occurring waters, return flows from other water rights, or other water sources not associated with the water right holder or sought by the applicant.

Nothing in this Model Agreement attempts to spell out in detail what steps might actually qualify as appropriate conservation measures. Such efforts as improved efficiency in manufacturing processes, the substitution of drip irrigation for sprinklers, or the introduction by a public water supply enterprise of requirement that customers use low flow toilets or showerheads would all be appropriate examples. The Model Agreement leaves the precise details regarding the suitability of these or other possible conservation measures to be developed by the regulatory and planning processes prescribed for the State Agency.

Cross-references: §1B-1-01 (purposes and scope of agreement.

§2B-2-04 DROUGHT

"Drought" conditions means conditions brought about by the lack of precipitation or water stored in the soil, in a quantity agreeable to the Parties, from the mean annual [rainfall] precipitation and water measured in soil.

Commentary: Management action will arise from a drought, or lack of mean annual rainfall, but could arise from other causes as well, such as the collapse of a dam with the resulting draining of a reservoir on which the Commission users depend. The definition should be determined, in large measure, by the use intended. Then a "drought management strategy" would be a specific course of conduct planned by the Commission as a necessary or appropriate response to the lack of precipitation.

Cross-references: §1B-1-01 (purpose and scope of agreement).

§2B-2-05 EQUITABLE AND REASONABLE UTILIZATION, CONSUMPTION OR DIVERSION

Utilization, consumption or diversion of a transboundary water resource in an equitable and reasonable manner requires taking into account all relevant factors and circumstances, including:

(a) Geographic, hydrographic, hydrological, climatic, ecological and other factors of a natural character;

(b) The social and economic needs of the Parties concerned;

(c) The population dependent on the water resource in each of the Parties;

(d) The effects of the use or uses of the water resources in by one Part on other Parties;

(e) Existing and potential uses of the water resource;

(f) Conservation, protection, development and economy of use of the water resource and the costs of measures taken to that effect;

(g) The availability of alternatives, of comparable value, to a particular planned or existing use.

(h) The potential or actual material injury or harm to other Parties utilizing the shared water resource.

Commentary. This definition is based in the 1997 *Convention on the Law of the Non-Navigational Uses of International Watercourses*, which was approved by the General Assembly of the United Nations by a vote of all but three nations. Drafters may wish to revert to the use of the term "equitable apportionment." However, this phraseology and definition are recommended since it is more encompassing and more descriptive of water-related activities that will affect water availability to the Parties.

Cross-references: §1B-1-02 (coordination and cooperation); §2B-2-02 (equitable and reasonable utilization, consumption or diversion); §4B-1-01 (exclusive jurisdiction and control).

§2B-2-06 FLOOD

"Flood" conditions means conditions resulting from heavy runoff with a frequency agreeable to the Parties.

Commentary: The flood condition is almost the opposite of a drought. A large amount of water is to be controlled by facilities of the Commission. The Parties are to agree as to the frequency of the flow of high waters in the Basin. Most of the time, these flows are during periods that exceed the amount of flow that occurs during the years of mean annual precipitation.

Cross-references: §1B-1-01 (purposes and scope of agreement); 4B-1-03 (water levels protected); §4B-1-05 (flood protection works).

§2B-2-07 PARTY OR PARTIES

"Party or Parties" means, unless the text otherwise indicates, those sovereign governments signatory to this Agreement.

Commentary: Defining the terms in this way avoids the need to include similar language at numerous points throughout the agreement.

Cross-references: §3B-1-01 (use of party officials); §3B-1-02 (substitution of officials); §3B-1-03 (implementation and verification of agreement); §3B-1-04 (funding); §3B-2-01 (rights in territory of other party); §3B-2-02 (storage and diversion); §3B-2-03 (eminent domain); §3B-2-04 (navigational servitude); §4B-1-01 (exclusive jurisdiction and control); §4B-1-02 (water allocation); §4B-1-03 (water levels protected); §4B-1-04 (underground water; limit on withdrawals); §4B-1-05 (flood protection works); §4B-1-06 (augmentation of supply); §4B-1-07 (water quality); §5B-1-02 (right to litigate).

§2C-2-08 UNDERGROUND WATER

The term "underground water" means water found beneath the ground, regardless of whether flowing through defined channels or percolating through the ground, whether the result of natural or artificial recharge.

Commentary: This definition of "underground water" includes all forms of water in the ground, being equivalent to terms such as "ground water," "groundwater," or similar expressions. It excludes soil (capillary) moisture which might be drawn upon by plants but cannot practically be withdrawn by direct human activity. *See* Rice & White, at 173. A somewhat more precise definition is found in the Illinois Water Use Act: water under the ground where the fluid pressure in the pore space is equal to or greater than atmospheric pressure. *See also* Dellapenna, § 6.04; Earl Finbar Murphy, *Quantitative Groundwater,* 3 Waters and Water Rights §18.02.

Cross-references: §1B-1-01 (purposes and scope of agreement); 2B-2-09 (waters of the basin); 4B-1-07 (water quality).

§2B-2-09 WATERS OF THE BASIN

"Waters of the Basin" shall include all water found within the Basin, whether surface, underground, or atmospheric water. Surface water is defined to include water in rivers, streams, lakes and reservoirs, and wetlands. Surface waters include all waters on the surface of the basin, whether in the form of rivers and streams, lakes and reservoirs, or wetlands. Underground water shall include any water extracted from underground sources, whether by well, natural spring, artesian flow or other method. The phrases "atmospheric water" means water produced by natural or artificial changes in the composition, motions, and resulting behavior of the atmosphere or clouds.

Commentary: This definition would be included to make it clear that underground water and atmospheric water are included within the scope of the agreement, if that is the intent of the Parties. The technological questions relating to atmospheric water may result in uncertainty regarding its allocation, but to the extent the Parties wish to reach a complete agreement, the matter should be addressed, or recognition given to the fact that the Parties have chosen to reserve that issue for later resolution. The Parties should also decide whether water imported from other basins should be included within the scope of the agreement. If it is not to be so included, that exclusion should be noted in this section.

Cross-references: §1B-1-01 (purposes and scope of agreement); §2B-2-01 (atmospheric water); §2B-2-08 (underground water); §4B-1-02 (water allocation); §4B-1-07 (water quality).

ARTICLE 3B

ADMINISTRATION

Part 1 Administration Officials

§3B-1-01 USE OF WATER MANAGEMENT OFFICIALS OF THE PARTIES

It shall be the duty of the Parties to administer this Agreement through the official of each Party who is now or may hereafter be charged with the duty of administering the public water supplies, and to collect and correlate through such officials the data necessary for the proper administration of the provisions of this Agreement. Such officials may, by unanimous action, adopt rules and regulations consistent with the provisions of this agreement.

Commentary: This section is one of two sections that provide a minimal means of administering the agreement. If a more structured or active administration is desired, a commission and authority may be established; *see* Model A (Coordination and Cooperation) and Model C (Comprehensive Management) for appropriate provisions and commentary.

Cross-reference: §1B-1-01 (purposes and scope of agreement); §1B-1-02 (coordination and cooperation); §1B-1-03 (good faith implementation); §2B-2-07 (party or parties); §3B-1-02 (substitution of officials); §3B-1-04 (funding).

Similar Agreements: *La Plata River Compact*, 43 Stat. 796 (1925); *Republican River Compact*, 57 Stat. 86 (1943); *Snake River Compact*, 64 Stat. 29 (1949); *Costilla Creek Compact*, 60 Stat. 246 (1946); amended 77 Stat. 350 (1963); *Upper Niobrara River Compact*, 83 Stat. 86 (1969).

§3B-1-02 SUBSTITUTION OF OFFICIALS

Whenever any official of any Party is designated to perform any duty under this agreement, such designation shall be interpreted to include the Party's official or officials upon whom the duties now performed by such official may hereafter devolve.

Commentary: This is the second of two sections that provide a minimal means of administering the agreement. Section 3-1-02 is included to guard against confusion in the event there is a subsequent reorganization of a party's government.

Cross References: §1B-1-01 (purposes and scope of agreement); §1B-1-02 (coordination and cooperation); §1B-1-03 (good faith implementation); §2B-2-02 (basin); §2B-2-07 (party or parties); §3B-1-01 (use of party officials); §3B-1-04 (funding).

Similar Agreements: *La Plata River Compact*, 43 Stat. 796 (1925); *South Platte River Compact*, 44 Stat. 195 (1923).

§3B-1-03 IMPLEMENTATION AND VERIFICATION OF AGREEMENT

(a) Each Party shall identify or maintain the administrative machinery necessary to implement the provisions of this Agreement, and, where several governmental institutions are involved, create the necessary co-ordinating mechanism for the authorities dealing with designated aspects of the environment.

(b) Each Party shall establish, maintain, and operate such suitable water gaging stations and facilities for measuring water quantity and quality as it finds necessary to administer and effect verification of this agreement.

Commentary: Implementation and verification of the Agreement will require administrative and technical support that must be provided by the Parties. This section obligates the Parties to provide that support.

Cross-references: §1B-1-01 (purposes and scope of agreement); §1B-1-02 (coordination and cooperation); §1B-1-03 (good faith implementation); §2B-2-07 (party or parties); §3B-1-04 (funding); §3B-2-01 (rights in territory of other party); §3B-2-02 (storage and diversion); §3B-2-03 (eminent domain); §3B-2-04 (navigational servitude); §4B-1-01 (exclusive jurisdiction and control); §4B-1-02 (water allocation); §4B-1-03 (water levels protected); §4B-1-04 (underground water; limit on withdrawals); §4B-1-05 (flood protection works); §4B-1-06 (augmentation of supply); §4B-1-07 (water quality).

Similar Agreements: *La Plata River Compact*, 43 Stat. 796 (1925).

§3B-1-04   FUNDING

Each Party shall allocate sufficient qualified personnel with adequate enforcement powers and sufficient funds to accomplish the tasks necessary for the implementation of this Agreement.

Commentary: In the case of simple allocation agreements in which no commission is established, funding provisions are not normally included. However, in order to ensure no misunderstanding exists concerning the responsibilities of each Party, an explicit provision may be preferable. Section 3B-1-04 is designed to avoid disputes over financing by requiring that each Party will operate the necessary facilities within its borders.

Cross References: §1B-1-01 (purposes and scope of agreement); §1B-1-02 (coordination and cooperation); §1B-1-03 (good faith implementation); §2B-2-07 (party or parties); §3B-1-03 (implementation and verification of agreement); §3B-2-01 (rights in territory of other party); §3B-2-02 (storage and diversion); §3B-2-03 (eminent domain); §3B-2-04 (navigational servitude); §4B-1-01 (exclusive jurisdiction and control); §4B-1-02 (water allocation); §4B-1-03 (water levels protected); §4B-1-04 (underground water; limit on withdrawals); §4B-1-05 (flood protection works); §4B-1-06 (augmentation of supply); §4B-1-07 (water quality).

Similar Agreements: ASEAN Agreement on the Conservation of Nature and Natural Resources, 1985.

Part 2   Activities Within the Territory of Other Party

§3B-2-01   RIGHTS IN TERRITORY OF OTHER PARTY (Optional)

Either Party shall have the right, in accordance with the laws of the other Party, to file applications for and obtain consent to construct or participate in the construction and use of any dam, storage reservoir, or diversion works in the territory of the other Party for the purpose of conserving and regulating the apportioned water without prejudice based on extra-territorial status; provided, that such right is subject to the rights of the other Party to control, regulate, and use water apportioned to it.

Commentary: To achieve efficient use of allocated water, it may be desirable for one Party or its citizens to construct reservoirs or other works within the boundaries of the other Party. This is the first of two sections that strive to gain concurrence and consent to do so, while preserving the rights of the party in which the works are constructed to control resource use within its territory.

Cross References: §1B-1-01 (purposes and scope of agreement); §1B-1-04 (preservation of federal rights); §1B-1-05 (national security); §2B-1-03 (consent to jurisdiction); §2B-2-07 (party or parties); §3B-1-03 (implementation and verification of agreement); §3B-2-02 (storage and diversion); §3B-2-03 (eminent domain); §3B-2-04 (navigational servitude); §4B-1-01 (exclusive jurisdiction and control).

Similar Agreements: *Belle Fourche River Compact*, 58 Stat. 94 (1944); *Republican River Compact*, 57 Stat. 86 (1943); *Snake River Compact*, 64 Stat. 29 (1949); *Upper Colorado River Basin Compact*, 63 Stat. 31 (1948); *Yellowstone River Compact,* 65 Stat. 663 (1950).

§3B-2-02   STORAGE AND DIVERSION (Optional)

Each claim hereafter initiated for storage or diversion of water in the territory of one Party for use by another Party shall be filed in the appropriate office of the Party in which the water is to be diverted, and a duplicate copy of the application including a map showing the character and location of the proposed facilities and the location(s) of the proposed uses shall be filed in the appropriate office of the Party from which the water is to be withdrawn. Any such construction or diversion by one Party within the territory of a second Party shall be subject to all appropriate laws and regulations of the second Party.

Commentary: This is the second of two sections that strive to gain concurrence and consent to do so, while preserving the rights of the party in which the works are constructed to control resource use within its territory.

Cross References: §1B-1-01 (purposes and scope of agreement); §1B-1-04 (preservation of federal rights); §1B-1-05 (national security); §2B-1-03 (consent to jurisdiction); §2B-2-07 (party or parties); §3B-1-03 (implementation and verification of agreement); §3B-2-01 (rights in territory of other party); §3B-2-03 (eminent domain); §3B-2-04 (navigational servitude); §4B-1-01 (exclusive jurisdiction and control).

Similar Agreements: *Belle Fourche River Compact*, 58 Stat. 94 (1944); *Snake River Compact*, 64 Stat. 29 (1949); *Yellowstone River Compact,* 65 Stat. 663 (1950).

§3B-2-03   EMINENT DOMAIN (Optional)

(a) Any Party, or person or other entity claiming water pursuant to the allocation of water to either Party, shall have the right to acquire necessary property rights in the territory of another Party by purchase or through the exercise of the power of eminent domain for the construction, operation and maintenance of storage reservoirs and of appurtenant works, canals, and conduits required for the enjoyment of the privileges granted by Article 4B; provided, however, that the Party, person, or entity exercising such rights shall pay to the political subdivisions of the Party in which such works are located, each and every year during which such rights are enjoyed for such purposes, a sum of money equivalent to the average annual amount of current year taxes assessed against the lands and improvements thereon during the years preceding the use of such lands in reimbursement for the loss of taxes to said political subdivision of the Party.

(b) Such power of condemnation shall be exercised in accordance with the provisions of any law applicable to the jurisdiction in which the property is located.

(c) Nothing in this Agreement authorizes the taking of any existing vested property right in the use of water except for just compensation, in accordance with the internal laws of the Party in which the property or usafructary right exists.

Commentary: In order to actually use the water allocated by the agreement, it may be necessary for one of the Parties or its citizens to construct reservoirs or other works within the boundaries of the other Party. This provision allows that to be done, through eminent domain if necessary, but also provides for payments in lieu of property taxes to avoid problems which might arise if one Party attempted to tax property belonging to another. If this provision is not included, the use of eminent domain will present a question of Party law in the Party in which the works are to be constructed. This alternative is likely to be adopted only within the United States or another federal system; issues of sovereignty may preclude use of this alternative as between sovereign states on the international level.

This provision expressly requires "just compensation" for any taking of property rights. The "just compensation" will, however, depend largely on the individual internal laws of the Parties themselves. In the United States, the recent rulings in regulatory by the Supreme Court have held that a serious impairment of the value of land by a regulation of its use must be compensated, but have noted that the State could diminish the value of a water right by as much as 95% without incurring liability, at least when a system of regulated riparian rights exist. *Lucas*, 112 S. Ct. at 2895 n. 8. *See also* J. Peter Byrne, *The Arguments for the Abolition of the Regulatory Takings Doctrine*, 22 Ecol. L.Q. 89 (1995); Oliver Houck, *Why Do We Protect Endangered Species, and What Does That Say about Whether Restrictions on Private Property to Protect Them Constitute "Takings"?*, 80 Iowa L. Rev. 297 (1995); Joseph Sax, *The Constitution, Property Rights and the Future of Water Law*, 61 U. Colo. L. Rev. 257 (1990).

Cross References: §1B-1-01 (purposes and scope of agreement); §1B-1-04 (preservation of federal rights); §1B-1-05 (national security); §2B-1-03 (consent to jurisdiction); §2B-2-07 (party or parties); §3B-1-03 (implementation and verification of agreement); §3B-2-01 (rights in territory of other party); §3B-2-02 (storage and diversion); §4B-1-01 (exclusive jurisdiction and control).

Similar Agreements: *Belle Fourche River Compact*, 58 Stat. 94 (1944); *Republican River Compact*, 57 Stat. 86 (1943); *Snake River Compact*, 64 Stat. 29 (1949); *Upper Colorado River Basin Compact*, 63 Stat. 31 (1948); *Yellowstone River Compact*, 65 Stat. 663 (1950).

§3B-2-04 NAVIGATION SERVITUDE (For international use)

(a) The Parties agree that the navigation of the _____ River shall forever continue free and open for the purposes of commerce to the inhabitants and to the ships, vessels, and boats of both Parties equally, subject, however, to any laws and regulations of either Party, within its own territory, not inconsistent with such privilege of free navigation and applying equally and without discrimination to the inhabitants, ships, vessels, and boats of both Parties.

(b) It is further agreed that so long as this Agreement shall remain in force, this same right of navigation shall extend to all waters bodies, tributaries and canals connecting boundary waters, now existing or which may hereafter be constructed on either side of the line. Either of the Parties may adopt rules and regulations governing the use of such connecting waters bodies, tributaries and canals within its own territory and may charge tolls for the use thereof, but all such rules and regulations and all tolls charged shall apply alike to the subjects or citizens of the Parties and the ships, aid vessels, and boats of both of the Parties, and they shall be placed on terms of equality in the use thereof.

Commentary: A clear statement of navigational interests of each party is set forth. Should some other paramount use between the Parties dominate at the time of treating/negotiations this would be set forth herein.

Cross-references: §1B-1-01 (purposes and scope of agreement); §1B-1-04 (preservation of federal rights); §1B-1-05 (national security); §2B-1-03 (consent to jurisdiction); §2B-2-07 (party or parties); §3B-1-03 (implementation and verification of agreement); §3B-2-01 (rights in territory of other party); §3B-2-02 (storage and diversion); §4B-1-01 (exclusive jurisdiction and control). §4B-1-03 (water levels protected);

Similar Agreements: *Treaty between the United States and Great Britain relating to Boundary Waters*, 36 Stat. 2451 (1909).

ARTICLE 4B

EQUITABLE AND REASONABLE USE OF WATER

§4B-1-01 EXCLUSIVE JURISDICTION AND CONTROL

(a) Each of the Parties reserves to itself, unless otherwise mandated by federal law or contractually agreed upon by the Parties, the exclusive control over the equitable and reasonable utilization, consumption or diversion of all waters within its borders.

(b) The Parties agree that any use or diversion from their natural channel of shared waters which result in injury to one Party's equitable and reasonable utilization of the shared waters shall give to the injured Party, or citizens of that Party, the same legal rights and entitlements to the same legal remedies as if such injury took place within the jurisdiction of the Party where such use or diversion occurs. Requirements for legal standing of citizens of the Party incurring injury by actions within the territory of the other Party shall be identical to those established for the citizens of the other Party.

(c) It is understood, however, that neither Party intends by the foregoing provision to surrender any right that it may have to enjoin or otherwise object to any interference with nor diversion by the other Party of shared waters that has a reasonable potential to cause material injury to the equitable and reasonable utilization of shared waters within its jurisdiction.

Commentary: This provision establishes the principle of the sovereign right of each Party to allocate or otherwise utilize and control the waters within its borders, constrained only by the the requirement that the such use be reasonable and equitable. Enforcement of this principle shall be based on those causes of actions and remedies available in tort within the legal system of the Party causing the injury. This section does provide for prospective relief from use, interference or diversion that may have a reasonable potential to cause material harm. A "reasonable potential" to cause material injury would be determined as a matter of law. The provision also resolves any legal standing issues that may arise in the case of citizens of one Party requesting legal intervention within the territory of the other Party.

Cross-references: §1B-1-01 (purposes and scope of agreement); §1B-1-04 (preservation of federal rights); §1B-1-05 (national security); §2B-1-03 (consent to jurisdiction); §2B-2-05 (equitable and reasonable utilization); §2B-2-07 (party or parties); §3B-1-03 (implementation and verification of agreement); §3B-2-01 (rights in territory of other party); §3B-2-02 (storage and diversion); §3B-2-03 (eminent domain); §3B-2-04 (navigational servitude); §4B-1-01 (exclusive jurisdiction and control).

Similar Agreements: *Treaty between the United States and Great Britain relating to Boundary Waters, and Questions arising between the United States and Canada*, 36 Stat. 2451 (1909); *Sabine River Compact*, 68 Stat. 690 (1953).

§4B-1-02 WATER ALLOCATION   (Optional)

[See Annex B for specific allocation alternatives]

Commentary: The basic allocation is a matter for negotiation. It may be based upon the relative geographic areas of the Parties, the relative contribution of each Party to the flow of the boundary stream, or any other method that is agreed upon as being equitable. A significant issue to be resolved in the negotiations is the means or methods used to verify compliance with the allocation provisions. A simple expedient of measuring flow at a specific point may be acceptable in cases where the river flow is consistently stable. However, in cases where extreme variations in flow occur or in geographic regions in which unstable meteorological and climatic changes are the norm, verification may require a sophisticated, complex scheme based on consistently measured consumption by the Parties.

The actual agreement may be relatively simple, as in the case of the *Sabine River Compact*, 68 Stat. 690 (1953), amended 76 Stat. 34 (1962), 91 Stat. 281 (1977), 106 Stat. 4661 (1992) between Texas and Louisiana. The Sabine originates in Texas, then forms the border between the two states. Texas is given the right to unrestricted use of the water above the gaging station at Logansport, where the river becomes the state boundary, except for an essentially de minimis minimum flow requirement. The water in the boundary reach is allocated equally between the two states, and any withdrawal from tributaries to that reach of the river are charged against the withdrawing state's allocation. A more complicated approach is seen in the *Red River Compact* among Texas, Oklahoma, Arkansas, and Louisiana, 94 Stat. 3305 (1978). The Red River serves not only as a state boundary, but also flows across borders. In addition, both appropriative and riparian water rights are recognized, depending on which state is involved. The resulting allocation is based on a division of the river into five reaches, with separate allocations for sub-basins within each reach.

Annex B provides a number of alternatives for allocation formulas that have been utilized in interstate compacts in the United States. (*See* Zachary McCormick, _____)

Cross-reference: §1B-1-01 (purposes and scope of agreement); §1B-1-04 (preservation of federal rights); §1B-1-05 (national security); §2B-1-03 (consent to jurisdiction); §2B-2-07 (party or parties); §2B-2-09 (waters of the basin); §3B-1-03 (implementation and verification of agreement); §3B-2-01 (rights in territory of other party); §3B-2-02 (storage and diversion); §3B-2-03 (eminent domain); §3B-2-04 (navigational servitude); §4B-1-02 (water allocation); §4B-1-03 (water levels protected); §4B-1-04 (underground water; limit on withdrawals); §4B-1-05 (flood protection works); §4B-1-06 (augmentation of supply); §4B-1-07 (water quality).

Similar Agreements: See Annex B

§4B-1-03 WATER LEVELS PROTECTED (Optional)

(a) The Parties agree that, except for the dams, reservoirs, obstructions, and diversions heretofore permitted or hereafter provided for by special agreement between the Parties hereto, no additional dams, reservoirs, obstructions or diversions of the shared waters of the _____ River Basin which affect the natural level or flow of boundary waters on the other side of the line shall be made except by approval of the other Party.

(b) The signatory Parties agree to furnish the other Party with complete documentation of any planned remedial or protective works or any dams, reservoirs or other diversions or obstructions to waters flowing into or from waters of the _____ River Basin. The other Party shall analyze the documentation to determine the potential flood damages that may therefrom arise and consult with the signatory Parties concerning the findings of such analysis.

(c) The foregoing provisions are not intended to limit or interfere with the existing rights of the Parties to undertake and carry on governmental works in shared waters for water development activities for economic growth, public health, recreational activities or environmental protection, provided that such works are wholly within its jurisdiction and do not materially affect the level or flow of the waters available to the other Party.

Commentary: A significant source of controversy develops during the construction and operation of water supply reservoirs and hydropower facilities. In both cases, significant reduction in flow may dramatically lower the flow in the shared water resource. Although this reduction may be limited to the initial start-up period and may be limited to a period of several years, severe economic and social impact may occur to other Parties. As importantly, releases from the works must be coordinated to assure downstream users are not affected. In the case of water supply reservoirs this may become critical during periods of drought. In the case of hydropower dams, especially those of a "peaking power" nature, the timing of release may be critical. This provision requires the sharing of data concerning such works and establishes a means of analysis of potential effects by specifically allowing for "special agreements" as regards raised levels above the natural level on transboundary rivers between the Parties. It also prohibits "dams or other obstructions" from raising the natural level of waters on the other side without such works having the approval of the other Party.

Cross-references: §1B-1-01 (purposes and scope of agreement); §2B-1-03 (consent to jurisdiction); §2B-2-02 (basin); §2B-2-06 (flood); §2B-2-07 (party or parties); §3B-1-03 (implementation and verification of agreement); §3B-2-01 (rights in territory of other party); §3B-2-02 (storage and diversion); §3B-2-04 (navigational servitude); §4B-1-01 (exclusive jurisdiction and control); §4B-1-02 (water allocation); §4B-1-03 (water levels protected); §4B-1-04 (underground water; limit on withdrawals); §4B-1-05 (flood protection works); §4B-1-06 (augmentation of supply); §4B-1-07 (water quality).

Similar Agreements: *Treaty between the United States and Great Britain relating to Boundary Waters*, 36 Stat. 2451 (1909); *Pecos River Compact* (1948).

§4B-1-04 UNDERGROUND WATER; LIMIT ON WITHDRAWALS (Optional)

When such action is necessary [to maintain an allocation set out elsewhere], the Parties shall regulate, in the same manner that surface flow is regulated, withdrawal of water from irrigation wells located within ___ miles of the river or its tributaries.

Commentary: If underground water is subject to the allocation provisions of the agreement, it may be useful to specifically address the steps to be taken with respect to withdrawals. This provision, adapted from the *Big Blue River Compact* between Kansas and Nebraska, 86 Stat. 193 (1972), uses a distance limitation to determine which wells fall within the scope of the agreement. If it is possible to establish the hydrological connection between all wells and the surface flow, the mileage limitation may be replaced with references to wells with such connection. In the absence of definitive hydrologic information, the mileage limitation may make administration easier, if less accurate.

No further specific allocation systems for underground water are provided because it is assumed that if underground water is allocated by agreement, that allocation will be in conjunction with allocation of related surface water sources and the allocation of underground water will be incorporated as part of the overall allocation of water. If underground water is allocated independently from surface water, the parties might use the surface models as a guide with respect to types of allocations (proportional, guaranteed minimum, etc.)

Cross References: §1B-1-01 (purposes and scope of agreement); §2B-1-03 (consent to jurisdiction); §2B-2-07 (party or parties); §3B-1-03 (implementation and verification of agreement); §3B-2-01 (rights in territory of other party); §3B-2-02 (storage and diversion); §3B-2-04 (navigational servitude); §4B-1-01 (exclusive jurisdiction and control); §4B-1-02 (water allocation); §4B-1-03 (water levels protected); §4B-1-05 (flood protection works); §4B-1-06 (augmentation of supply); §4B-1-07 (water quality).

Similar Agreements: *Kansas-Nebraska Big Blue River Compact*, 86 Stat. 193 (1972).

§4B-1-05 FLOOD PROTECTION WORKS (Optional)

(a) As a general concept, the use of the channels of the waters of the _____ River Basin for the discharge of flood or other excess waters shall be free and not subject to limitation by either country, and neither country shall have any claim against the other in respect of any damage caused by such use. However, the signatory Parties declare their intention to manage flood control programs and activities in such manner, consistent with the normal operations of its hydraulic systems, as to avoid, as far as feasible, material damage in the territory of the other.

(b) Each Party agrees to furnish the other Party with complete documentation of existing and planned flood protection programs and works. The other Party shall analyze the documentation to determine the potential flood damages that may therefrom arise and enter into consultations and negotiations as necessary concerning the findings of such analysis.

Commentary: Flood control policies and works can have a dramatic effect on the timing and elevation of water levels and thus may become a major contentious issue between the Parties. The issue often will transcend "equitable and reasonable utilization" and should be addressed as an individual area of coordination. This provision recognizes the sovereign right of each Party to make efforts to safeguard its people and economic forces from flood damages but establishes an avenue for the sharing of data on flood control efforts as well as an independent analysis of the effects of those efforts on other Parties.

Cross-references: §1B-1-01 (purposes and scope of agreement); §1B-1-04 (preservation of federal rights); §1B-1-05 (national security); §2B-1-03 (consent to jurisdiction); §2B-2-06 (flood); §2B-2-07 (party or parties); §3B-1-03 (implementation and verification of agreement); §3B-2-01 (rights in territory of other party); §3B-2-02 (storage and diversion); §3B-2-03 (eminent domain); §3B-2-04 (navigational servitude); §4B-1-01 (exclusive jurisdiction and control); §4B-1-02 (water allocation); §4B-1-03 (water levels protected); §4B-1-04 (underground water; limit on withdrawals); §4B-1-06 (augmentation of supply); §4B-1-07 (water quality).

Similar Agreements: *Treaty between the United States of America and Mexico, Utilization of Waters of the Colorado and Tijuana rivers and of the Rio Grande*, 59 Stat. 1219 (1945).

§4B-1-06 AUGMENTATION OF SUPPLY (Optional)

(a) Any importation of water from outside the Basin shall be excluded from the provisions set forth elsewhere in this agreement, and the Party importing such water shall have the right to full and complete use and consumption of such imported water.

(b) Any Party which augments precipitation within the Basin shall be entitled to full and exclusive use of additional water supplies resulting from such augmentation, notwithstanding any other standard of allocation set forth in this agreement. In the event the Parties cannot agree on whether or to what extent precipitation has been augmented, the Party asserting the right to such increased supplies shall bear the burden of proving that the increase, if any, was the result of the Party's augmentation efforts and not simply the result of natural variation in precipitation amounts.

(c) Any Party implementing a conservation program with respect to water supplies shall be entitled to full and complete use and consumption of all increased supplies resulting from such conservation program. The burden of showing such increase shall rest on the Party claiming such increase.

Commentary: Section 4B-1-06 makes clear that if a Party arranges to increase supplies by importing water, it need not share those additional supplies. Section 4B-1-06(b) applies the same principle for precipitation augmentation. Section 4B-1-06(c) provides encouragement for conservation by rewarding the Party that undertakes that effort. Caution should be exercised in incorporating this provision, however, inasmuch as the level of conservation efforts between the Parties may be unequal at the time the agreement is negotiated. A Party that has already made significant efforts should not be placed at a disadvantage relative to a Party that, prior to the agreement, made little effort to conserve.

Cross References: §1B-1-01 (purposes and scope of agreement); §2B-2-07 (party or parties); §3B-2-01 (rights in territory of other party); §3B-2-02 (storage and diversion); §3B-2-03 (eminent domain); §4B-1-01 (exclusive jurisdiction and control); §4B-1-02 (water allocation); §4B-1-03 (water levels protected); §4B-1-04 (underground water; limit on withdrawals); §4B-1-05 (flood protection works); §4B-1-07 (water quality).

Similar Agreements: Rio Grande Compact, 53 Stat. 785 (1938); *Pecos River Compact* (1948).

§4B-1-07 WATER QUALITY (Optional)

<u>Alternative 1</u> (For U.S. use)

The Parties shall:

(a) **Manage the waters of the Basin within their jurisdiction to maintain ecosystem integrity, preserve and protect aquatic ecosystems effectively from any form of (significant) degradation on a drainage Basin or sub-Basin basis. Natural water quality solutions, such as riparian vegetated buffers along the _____ River and its tributaries, will be utilized to the maximum extent possible.**

(b) **Publish biological, health, physical and chemical quality criteria for all water bodies (surface and underground water), according to Basin capacities and needs, with a view to an ongoing improvement of water quality.**

(c) **Establish standards for the discharge of effluents and for the receiving waters, no less stringent than the effluent limitations established by the U.S. Environmental Protection Agency, including standards for land use management.**

(d) **Establish minimum flow criteria to insure nourishment of wetlands and riparian buffers as necessary to properly filter nitrates and phosphorous arising from nonpoint runoff.**

(e) **Maintain the quality of the Waters of the Basin at or above water quality standards as may be adopted, now or hereafter, by the water pollution control** agencies of the respective Parties in compliance with the provisions of the Clean Water Act, 33 U.S.C. §§1251 *et seq.*, and amendments thereto.

Alternative 2 (For international use)

The Parties mutually agree to:

(a) Comply with the principle of individual Party efforts to control natural and man-made water pollution within each Party and to the continuing support of both Parties in active water pollution control programs.

(b) Cooperate, through their appropriate Party agencies, in the investigation, abatement, and control of sources of alleged interparty pollution within the Basin.

(c) Cooperate in maintaining the quality of the Waters of the Basin at or above water quality standards as may be developed and agreed to by the Parties.

Commentary. The quality of the water that is allocated is as important as the quantity of water that is allocated. Poor quality water imposes risks that the Parties should consider. First there is the health risk to the population that uses the water for domestic purposes. Second if the available water will not meet the standards for certain industrial purposes, there is the risk that economic growth will be impaired. Finally there is the risk that quality degradation will have a severe impact on the ecology of the Basin, resulting in long-term sustainability complications. Integration of water quality and quantity is essential. Agenda 21 obligated all signatories to develop a program of water and sustainable development; *see* also A. Satre Ahlander, *Environmental Policies in the former Soviet Union*, Economic Policies for Sustainable Development, (Thomas Sterner, ed.), 1994) The UN *Convention on the Law of the Non-Navigational Uses of International Watercourses*, United Nations Document A/51/869 (1998), establishes the criterion that "(w)atercourse States shall, individually and, where appropriate, jointly, protect and preserve the ecosystems of international watercourses."

Surface and underground water may be degraded by a variety of factors. Major problems affecting the quality of these water resources arise, for instance, from inadequate domestic sewage treatment, inadequate controls on the discharge of industrial waste and effluent, the diversion of waters resulting in insufficient water to assimilate waste, the loss and destruction of catchment areas, the improper siting of industrial plants, deforestation and poor agricultural practices which cause leaching of nutrients and pesticides. Transboundary water sharing must include effective plans and programs that eliminate, or at least minimize, the possible sources of water quality degradation.

The complex interconnected nature of freshwater systems suggests that freshwater management should be systemically integrated, taking a catchment management approach which balances the needs of people and the environment. The Parties should manage the waters of the Basin preserve aquatic ecosystems, and protect them effectively from any form of degradation on a drainage Basin or sub-Basin basis. Scientific research has shown that the use of riparian vegetative buffers can have an extremely salutary effect on reducing pollution, especially sediments, nitrates and phosphorous. Such natural solutions for quality control should be utilized to the maximum possible extent. Periodic flushing flows at periodic intervals may be an effective technique to quickly rejuvenate degraded waters.

The Parties should establish biological, health, physical and chemical quality criteria for all significant water bodies in the Basin to continually improve water quality. The Parties should establish minimum standards both for discharge of effluents and for receiving waters. We recommend the Parties institute standards for land use management such as limits on agrochemical use, deforestation, and wasteful irrigation practices. Such rational land use standards should prevent land degradation, erosion and siltation of lakes and other water bodies.

Cross-references: §1B-1-01 (purposes and scope of agreement); §1B-1-04 (preservation of federal rights); §1B-1-05 (national security); §2B-1-05 (limited applicability); §2B-1-06 (annexes); §2B-2-07 (party or parties); §2B-2-08 (underground water); §2B-2-09 (waters of the basin);
§3B-1-03 (implementation and verification of agreement); §3B-2-01 (rights in territory of other party); §3B-2-02 (storage and diversion); §3B-2-03 (eminent domain); §3B-2-04 (navigational servitude); §4B-1-01 (exclusive jurisdiction and control); §4B-1-02 (water allocation); §4B-1-03 (water levels protected); §4B-1-04 (underground water; limit on withdrawals); §4B-1-05 (flood protection works); §4B-1-06 (augmentation of supply); §4B-1-07 (water quality).

Similar Agreements: Rio Grande Compact, 53 Stat. 785 (1939); *Klamath River Basin Compact*, 71 Stat. 497 (1957); Arkansas River Basin Compact of 1965, 80 Stat. 1409 (1966); Big Blue River Basin Compact, 86 Stat. 193 (1972); Arkansas River Basin Compact of 1970, 87 Stat. 569 (1973); *Delaware River Basin Compact*, Pub. L. 87-328, 75 Stat. 688 (1961); *Susquehanna River Basin Compact*, Pub. L. No. 91-575, 84 Stat. 1509 (1970); *Kansas-Nebraska Big Blue River Compact*, 86 Stat. 193 (1972); *Red River Compact*, 94 Stat. 3305 (1980); Oregon-California Goose Lake Interstate Compact, 98 Stat. 291 (1984); *Treaty between the United States and Great Britain relating to Boundary Waters, and Questions arising between the United States and Canada*, 36 Stat. 2451 (1909); Agreement between the United States and Canada on Great Lakes Water Quality, 1153 UNTS 187 (1978); *Convention on the Protection and Use of Transboundary Watercourses and International Lakes*, 31 I.L.M. 1312 (1992); *Convention on the Law of the Non-Navigational Uses of International Watercourses*, United Nations Document A/51/869 (1998).

ARTICLE 5B

DISPUTE RESOLUTION

(OPTIONAL)

Commentary. Disputes will inevitably arise as the Agreement is implemented. Thesed disputes may involve differences in interpretation of the Agreement's provisions or non-compliance with the Agreement itself. The disputes may also arise because of changing conditions that alter the effectiveness of the Agreement for one or more of the Parties. While a speedy and equitable process of dispute resolution serves all Parties well, some sovereign entities do not wish to enter into an obligatory process. In such a case, Article 5A may be omitted. In other instances, the Parties may recognize the need to institutionalize a dispute resolution process

§5A-1-01 RESOLUTION BY SIGNATORY PARTIES

Whenever any difference or dispute may arise between two or more Parties to this Agreement regarding any matters covered by this Agreement, particularly as to the interpretations of the Agreement and the legal rights of the parties, the Parties shall first make every effort to resolve the issue through negotiations and consultations based on the powers and duties herein described.

Commentary. This alternative is the least restrictive upon the sovereignty of the Parties yet expresses their recognition of the need for peaceful resolution of dipustes. The alternative is appropriate in those cases where no commission has been established to manage the data exchange.

Cross-references: §1A-1-01 (general policies); §1A-1-02 (coordination and cooperation); §1A-1-03 (good faith implementation); §1A-1-04 (preservation of federal rights); §1A-1-05 (national security); §2A-1-05 (powers of sovereign parties; withdrawal); §2A-1-06 (existing agencies); §2A-1-07 (limited applicability); §2A-2-05 (party or parties); §5A-1-02 (alternative 1, right to litigate).

Similar Agreements: *Agreement on Cooperation for the Sustainable Development of the Mekong River Basin*, 34 ILM 864 (1995).

§5A-1-02 RIGHT TO LITIGATE

Nothing in this agreement shall be construed to limit or prevent either Party from instituting or maintaining any action or proceeding, legal or equitable, in any tribunal of competent jurisdiction for the protection of any right under this agreement or the enforcement of any of its provisions.

Commentary: The existence of an appropriate tribunal may pose a problem in cases not involving an entity like the United States or European Union. It may be advisable to specify the tribunal in the agreement itself to avoid dispute over jurisdictional questions at a later date.

Cross References: §1A-1-01 (general policies); §1A-1-02 (coordination and cooperation); §1A-1-03 (good faith implementation); §1A-1-04 (preservation of federal rights); §1A-1-05 (national security); §2A-1-05 (powers of sovereign parties; withdrawal); §2A-1-06 (existing agencies); §2A-1-07 (limited applicability); §2A-2-05 (party or parties); §5A-1-01 (alternative 1, resolution by signatory parties,).

Similar Agreements: *Belle Fourche River Compact*, 1944; *Colorado River Compact*, 1928; *Snake River Compact*, 1949.

SIGNATURES

IN WITNESS WHEREOF, and in evidence of the adoption and enactment into law of this Agreement by the signatory Parties, the representatives of the sovereign States of _____, _____, _____ do hereby, in accordance with authority conferred by law, sign this Agreement in (__) duplicate original copies, as attested by the appropriate authorities of the respective sovereign Parties, and have caused the seals of the respective States to be hereunto affixed this ____ day of _____.

ANNEX B-1: ALLOCATION ALTERNATIVES

§4B-1-02 WATER ALLOCATION

<u>Alternative 1</u>: **PERCENTAGE ALLOCATION**:

(a) The Parties agree that the unappropriated Waters of the Basin as of the date of this agreement shall be allocated to each Party as follows:

___ % to [Upstream Party]
___ % to [Downstream Party]

(b) For storage of its allocated water, either Party shall have the right to purchase at cost up to ___% of the total storage capacity of any reservoir or reservoirs constructed by the other Party in its territory for irrigation of lands, or may construct reservoirs itself for the purpose of utilizing such water. Either Party may temporarily divert, or store for beneficial use, any unused part of the above percentages allotted to the other, but no continuing right shall be established thereby.

(c) Rights to the use of the Waters of the Basin, whether based on direct diversion or storage, are hereby recognized as of the date of this agreement to the extent these rights are valid under the law of the Party in which the use is made, and shall remain unimpaired hereby. These rights, together with the additional allocations made under this section, are agreed to be an equitable apportionment between the Parties of the Waters of the Basin.

(d) The waters allocated under this section and the other rights recognized under Article 4B are hereinafter referred to collectively as the apportioned water. For the purposes of the administration of this agreement and determining the apportioned water at any given date within a given calendar year, there shall be taken the sum of:

(1) The quantity of water in acre-feet [cubic meters] that passed the boundary lines of the Parties during the period _____;

(2) The quantity of water in acre-feet [cubic meters] in storage on that date in all reservoirs built in [Upstream Party] in the Basin subsequent to the date of this agreement; and

(3) The quantity of underground water withdrawn from aquifers tributary to surface water sources in the Basin during the period _____.

Commentary: Section 4B-1-02(a) is a simple proportional allocation for use in upstream/downstream conflicts. Both Parties share the risk of shortage in proportion to their percentage of allocation. Determination of the percentages is a matter of negotiation. It may be based upon amount of irrigable land, historic use, population, or some other factor, or it may simply be derived by bargaining.

In §4B-1-02(b) the right of the upstream Party to use reservoirs constructed by the downstream Party is similarly a matter for negotiation, but may result in the construction of fewer reservoirs than would otherwise be necessary.

Section 4B-1-02(c) excludes existing uses from the allocation. Inclusion of such a grandfather clause may avoid claims that water rights or other property interests have been taken as a result of the allocations. Inclusion of such a provision is reasonable only if there remains unallocated supply within the Basin. If only new uses are affected by the allocation, the actual proportion of the river's flow available to each Party may be different than the allocation for future use. This would be the case where one Party is using considerably more water than the other at the time the allocation is agreed upon.

Section 4B-1-02(d), or something similar, should be included, as there must be some quantity against which the percentages can be applied. In this model, taken from the *Belle Fourche Compact*, 58 Stat. 94 (1944), the state line is the measuring point, and a particular time period is used.

The state line may not, however, be appropriate because the upstream party can divert water before it reaches that point. It may be necessary to specify the use of gaging stations at particular points upstream of the relevant boundary, as is the case with the *Rio Grande Compact*, 53 Stat. 785 (1938). The *Rio Grande Compact* employs a set of tables to be used in calculating delivery obligations. The amount required to be delivered by the states at specified locations is determined by the flow measured at those locations. Where there are tributaries of significance, as on the upper Rio Grande, it may be necessary to specify specific locations on each tributary.

If underground water is included in the allocation, provision must be made to account for underground water withdrawals and availability. No specific mention is made of atmospheric water or water imported from other basins. In practice, this would mean that additional water from those sources would be subject to the allocation, if it were to enter into the flow of the river, because it would pass the gaging station. If such water is not to be included in the allocation, specific exclusions should be incorporated in the agreement to avoid future disputes over whether underground water was intended to be subject to the allocation system.

Cross References: See main text.

Similar Agreements: *Belle Fourche River Compact*, 58 Stat. 94 (1944); *Snake River Compact*, 64 Stat. 29 (1949); *Yellowstone River Compact,* 65 Stat. 663 (1950).

<u>**Alternative 2**:</u> **PROPORTIONATE ALLOCATION**

      (a) The waters of the _____ River are hereby equitably apportioned between the Parties as follows:

(1) At all times between the ___ day of [month] and the ___ day of the succeeding [month], each Party shall have the unrestricted right to the use of all water which may flow within its boundaries;

(2) By reason of the usual annual rise and fall, the flow of said river between the ___ day of [month] and the ___ day of [month] of each year, shall be apportioned between the Parties in the following manner:

(A) Each Party shall have the unrestricted right to use all the waters within its boundaries on each day when the mean daily flow at the [Downstream] station is ___ cubic feet [meters] per second, or more;

(B) On all other days [Upstream Party] shall deliver at the [Downstream] Gaging Station a quantity of water equivalent to ___ per cent of the mean flow at the [Upstream] Gaging Station for the preceding day, but not to exceed ___ cubic feet [meters] per second;

(3) Whenever the flow of the river is so low that in the judgment of the Parties by their designated officials, the greatest beneficial use of its waters may be secured by distributing all of its waters successively to the lands in each Party in alternating periods, in lieu of delivery of water as provided in the second paragraph of this section, the use of the waters may be so rotated between the two Parties in such manner, for such periods, and to continue for such time as the Parties may jointly determine;

(4) [Downstream Party] shall not at any time be entitled to receive nor shall [Upstream Party] be required to deliver any water not then necessary for beneficial use in [Downstream Party];

(5) A substantial delivery of water under the terms of this article shall be deemed a compliance with its provisions and minor and compensating irregularities in flow or delivery shall be disregarded.

(b) Verification of the allocation of waters through the use of gaging stations established by the Parties.

(1) [Upstream Party], at its own expense, shall establish and maintain two permanent stream-gaging stations upon the ___ River for the purpose of measuring and recording its flow, which shall be known as the [Upstream] Gaging Station and the [Downstream] Gaging Station, respectively.

(2) The [Upstream] Gaging Station shall be located at some convenient place near [geographic reference]. Suitable devices for ascertaining and recording the volume of all diversions from the river above Station, shall be established and maintained (without expense to [Downstream Party]), and whenever in this agreement reference is made to the flow of the river at [Upstream] Gaging Station,

it shall be construed to include the amount of the concurrent diversions and underground water withdrawals above said station.

(3) The [Downstream] Gaging Station shall be located at some convenient place within one mile of the boundary line between [Upstream Party] and [Downstream Party]. Whenever in this agreement reference is made to the flow of the river at the [Downstream] Gaging Station, it shall be construed to include the volume of any other water that may hereafter be diverted from said river or underground water tributary to said river in [Upstream Party] for use in [Downstream Party].

(4) Each of said stations shall be equipped with suitable devices for recording the flow of water in said river at all times between the ____ day of _____ and the ____ day of _____ of each year. The designated officials of the Parties shall make provision for cooperative gaging at the two stations, for the details of the operation, exchange of records and data, and publication of the facts.

Commentary: Alternative 2, adapted from the *La Plata Compact,* calls for apportionment only during certain periods of the year and only at times when the flow of the river falls below a certain level. It is directed towards upstream/downstream conflicts. La Plata is a relatively small river whose flow can fluctuate widely. Allocation was deemed necessary only in times of low flow. The state engineers of Colorado (the upstream state) and New Mexico (the downstream state) are given some power under the compact to coordinate the use of whatever water is available during those low-flow periods. This alternative provides for the actual allocation of the water when the triggering event has occurred. It contains an additional provision to allow for more efficient use of the water by allowing the Parties to allocate the entire flow to each Party in rotation. Finally, a de minimis clause is included to avoid argument over relatively minor problems. The alternative specifies how flows are to be calculated and who bears the cost of the gaging stations. It provides for adjustments of the measured amounts to take into account diversions and underground water withdrawals not recorded at the gaging stations. These diversions and underground water withdrawals are in effect charged to the Party that benefits from them.

Cross References: See main text.

Similar Agreements: *La Plata River Compact*, 43 Stat. 796 (1925).

Alternative 3: GUARANTEED MINIMUM QUANTITIES APPORTIONED.

(a) There is hereby apportioned from the _____ River to [Downstream Party] the exclusive reasonable use of [_____] acre feet [cubic meters] per annum, measured at [geographic location] which shall include all water necessary for the supply of any rights which may now exist.

(b) [Upstream Party] shall not cause the flow of the river at [geographic location] to fall below an aggregate of [ten times the quantity above] for any period of [ten] consecutive years determined in progressive continuing series, beginning with the ___ day of _____ next succeeding the effective date of this agreement.

Commentary: This alternative is derived from part of the *Colorado River Compact* of 1922. It is directed to upstream/downstream conflicts. The *Colorado River Compact* involves seven states, and the water is allocated not between states, but between the upper and lower basins of the river. The agreement calls for the lower basin to receive 7,500,000 acre feet per year, but in apparent recognition of the variable nature of the Colorado's flow, the compact also provides that the states of the upper basin will not cause the flow at Lee Ferry (the gaging station) to fall below 75,000,000 acre feet in any 10 year period. The risk of shortage is therefore on the states of the Upper Basin, but shortages in one year can be made up in another so long as the ten-year requirement is met. In this example the provision might read

(a) There is hereby apportioned from the Colorado River to Arizona the exclusive reasonable use of 7,500,000 acre feet per annum, measured at Lee Ferry which shall include all water necessary for the supply of any rights which may now exist.

(b) Nevada shall not cause the flow of the river at Lee Ferry to fall below an aggregate of 75.000,000 for any period of ten consecutive years determined in progressive continuing series, beginning with the 1st day of January next succeeding the effective date of this agreement.

This provision places the risk of shortfall on the upstream state, which is bound to deliver a specified amount regardless of changes in supply. A rolling average is used to determine compliance. The Colorado Compact uses a ten-year period, but the actual number is a matter for negotiation. While it is not absolutely necessary that some averaging period be chosen, the potential for variation in the flow of most rivers makes it a reasonable provision.

Cross References: See main text.

Similar Treaties and Agreements: *South Platte River Compact*, 44 Stat. 195 (1923); *Colorado River Compact*, approved 45 Stat. 1057 (1928).

Alternative 4: GUARANTEED QUANTITY; VESTED RIGHTS PROTECTED
(For Appropriative Rights states only)

(a1) Between the ___ day of [month] and the ___ day of [month] in each year, [Upstream Party] shall not permit diversions from the river to supply [Upstream Party] appropriations with a priority date subsequent to [date] to the extent that such diversions will diminish the flow of the river at the interstate gaging station below a mean flow of _____ cfs, except as limited by §4B-1-02(d).

(b) [Upstream Party] and [Downstream Party], at their joint expense, shall maintain a stream gaging station upon the _____ River near [geographic reference] for the purpose of measuring the amount of water flowing in said river from [Upstream Party] into [Downstream Party]. The location of said station may be changed from year to year as the river channels and flow conditions of said river may require.

(c) The officials responsible for administering public water supplies in each of the Parties shall make provision for the cooperative gaging at and the details of operation of said gaging station and for the exchange and publication of records and data.

(d) [Downstream Party] shall not be entitled to receive, and [Upstream Party] shall not be required to deliver, on any day any part of the flow of the river not then necessary for Beneficial Use in [Downstream Party]; provided, however, that a minimum of _____ cfs [cubic meters per second] shall be released, if available, to maintain minimum flows.

Commentary: This alternative is adapted from the *South Platte River Compact*, 44 Stat. 25 (1923), between Colorado and Nebraska. The provision is applicable only to those states within which Appropriative Rights law exists. The South Platte agreement provides for a minimum flow at the state boundary, but Colorado (the upstream state) did not actually guarantee that the flow would equal the minimum. Instead, Colorado agreed that it would not permit any diversions with a priority date later than June 14, 1897, if those diversions would reduce the flow at the boundary below 120 cfs. Under the compact, Nebraska would bear the risk that the river would have insufficient flow to meet the demands of Colorado appropriators with a priority date earlier than June 14, 1897, and still provide 120 cfs at the border. The result is a sharing of the risk of shortage. Nebraska bears the risk up to a certain point; thereafter, the risk of shortage falls on later Colorado appropriators.

This alternative seeks to accomplish the same result. It provides for a minimum flow at the Party boundary, but limits the obligation of the Upstream Party to provide that water by requiring only that the Upstream Party not permit diversions with a priority after a given date, if those diversions would reduce the flow below the minimum called for. Senior appropriators, whose rights presumably vested prior to the date of the compact, are still able to withdraw water even if those withdrawals result in a boundary line flow below the minimum. The Downstream Party thus bears the risk that the available water will be less than those vested rights plus the minimum.

Section 4B-1-02(d) is included to insure that water which can be used in the Upstream Party is not required to pass the border between the Parties unless it will be put to beneficial use in the Downstream Party. The use of the term "Beneficial Use" is restricted to those jurisdictions which adhere to the Law of Appropriate Rights and will determine whether the Downstream Party is entitled to release of water for purposes such as instream use in those jurisdictions. Minimum flows may be maintained by virtue of the provison. In addition, the beneficial use is referred to as "in" the Downstream Party. This could preclude a requirement for water to be released to satisfy the demands of water transfers (water markets) if those markets resulted in transfers outside the boundaries of the Downstream Party.

Cross References: See main text.

Similar Agreements: *South Platte River Compact*, 44 Stat. 195 (1923).

Alternative 5: STORAGE ALLOCATIONS.

(a) No reservoir or other storage facility, whether surface or underground, built after the date of this agreement solely to utilize the water allocated to either Party shall have a capacity in excess of _____ acre-feet [cubic meters], without the approval of both Parties.

(b) [Upstream Party] shall have free and unrestricted use of all waters originating in the Basin above [name] dam.

(c) [Downstream Party] shall have free and unrestricted use of all waters originating in the basin below [name] dam.

Commentary: If the Upstream Party builds excess storage capacity, the Downstream Party may not receive the full benefit of its bargain. If that is a concern, a storage limitation such as this can be included. Underground storage is within the scope of this provision, to allow for the possibility of storage in other than surface reservoirs.

In some cases, the Parties may determine that limitations on storage will satisfy any need for allocation. Storage is relatively easy to measure, and such agreements have the virtue of simplicity. Careful attention must be paid to definitions. The latter two sections of this article are derived from the *Canadian River Compact*, 66 Stat. 74 (1952). This agreement between New Mexico, Texas, and Oklahoma eventually required interpretation by the United States Supreme Court in Oklahoma *and Texas v. New Mexico*, 111 S. Ct. 2281 (1991). The case required the Court to decide the meaning of the term "originating"; more specifically, a key issue was whether water that entered the river above Conchas Dam but was then spilled or released from the dam "originated" above or below the dam. The Court found that this water originated below the dam. Even seemingly unambiguous terms such as "originating" must be used with care.

Cross References: See main text.

Similar Agreements: *Kansas-Nebraska Big Blue River Compact*, 86 Stat. 193, §5.2 (c) (1972); *Upper Niobrara River Compact*, 83 Stat. 86, Art. V (1969); *Canadian River Compact*, 66 Stat. 74, Art. IV (1952).

COMPREHENSIVE MANAGEMENT AGREEMENT RELATING TO THE SHARED USE OF TRANSBOUNDARY WATER RESOURCES

Model C

Contents

Comprehensive Management Agreement Relating to the Shared Use of Transboundary Water Resources

Introduction 84
Article 1C – Declaration of Policies and Purposes 86
General Policies 86
Purposes and Scope of Agreement 88
Objectives of Agreement (Optional) 90
Coordination and Cooperation 91
Preservation of Federal Rights (For U.S. use only) 93
National Security (For international use) 94
Article 2C – General Provisions 95
Part I: General Obligations 95
Effective Date 95
Consent of Jurisdiction (For U.S. use) 96
Duration of Agreement (Optional) 97
Powers of Sovereign Parties; Withdrawal (Optional, International use only) 98
Amendments and Supplements (Optional) 99
Existing Agencies 99
Severability 100
Annexes 100
Part II: Definitions 101
Atmospheric Water 101
Basin 101
Comprehensive Water Management Plan 102
Conservation Measures 102
Drought 103
Equitable and Reasonable Utilization 104
Flood 104
Interbasin Transfer 105
Party or Parties 105
Underground Water 106

Waters of the Basin 106
Article 3C – Administration...107
Part I: Administrative Authority..107
Commission Created... 107
Jurisdiction of the Commission 108
Commissioners 109
Status, Immunities and Privileges (Optional)... 110
Commission Organization and Staffing.. 112
Rules of Procedures 113
Commission Administration 113
Part II: Powers and Duties..115
General Powers and Duties... 115
Powers and Duties Reserved to the Commissioner 116
Powers and Duties to the Commission (Optional).. 118
Regulations: Enforcement... 119
Prohibited Activities (Optional).. 120
Commission Approval of Water Resources Project (Optional).. 121
Advisory Committees (Optional).. 122
Reports (Optional) 122
Condemnation Proceedings (Optional)... 123
Meetings, Hearings, and Records (Optional)... 125
Tort Liability... 126
Part III: Intergovernmental Relations...126
Coordination and Cooperation.. 126
Project Costs and Evaluation Criteria... 127
Projects of the Signatory Parties... 128
Cooperative Services 129

Article 4C – Comprehensive Water Management 130
Part I: Comprehensive Water Management 130
Joint Exercises of Sovereignty 130
Interrelationship of Water Resources 131
Comprehensive Water Management Plan 131
Purpose and Objectives of Comprehensive Water Management Plan 133
Conditions of Comprehensive Water Management Plan (Optional) 135
Deviation from Comprehensive Water Management Plan 136
Allocation During Flood Conditions 138
Allocation Under Drought Conditions 140
Non-impairment of Comprehensive Water Management Plan by State Action (Optional) ... 141
Part II: Water Allocation 142
Water Allocation, Generally 142
Waters not Subject to Allocation 143
Assessment and Enhancement of Basin Water Sources 144
Assessment and Classification of Basin Water Demands and Needs 145
Criteria for Water Allocation (Optional) 148
Watershed Management (Optional) 150
Part III: General Provisions 152
Existing Rights Recognized 152
Flood Protection Works 152
Minimum Flows 153
Withdrawals and Diversion; Protected Areas (Optional) 154
Augmentation of Supply 155
Water Quality 156
Underground Water; Limit on Withdrawals (Optional) 159
Atmospheric Water 160
Recreation (Optional) 161

Article 5C – Financing 162

Article 6C – Dispute Resolution 163
Good Faith Implementation... 164
Modification of Agreement.. 165
Material Breach 166
Negotiations and Consultations 167
Alternative 1 (Right to Litigate) 168
Right to Litigate (Optional) 168
Alternative 2 (Binding Alternative Dispute Resolution)................................... 168
Alternative Dispute Resolution.. 168
Conciliation and Mediation 170
Arbitration.. 171

Signatures 173

ANNEX C-1 – Financing (For Financing Alternative 2)................................... 174

ANNEX C-2 – Arbitral Panel 178

Index 181

INTRODUCTION

Model C, the Comprehensive Water Management Agreement, provides a model for comprehensive planning and management of shared water resources. This Model Agreement is based on the concept that the most efficient and effective allocation of shared water resources can be achieved only through management on a watershed basis, that the water resources and the associated riparian lands are an integrated whole whose interrelation requires integrated management to achieve optimal use especially during periods of drought. Such integrated management requires creation of a separate management entity to which the Parties will cede specified decision-making authority.

The agreement is extensive and considers all aspects of the management of the water resources of a particular basin. In addition to seeking optimal use of the waters of the particular basin, the objectives of this comprehensive, integrated agreement seeks to achieve an equitable allocation among all Parties to the agreement. It proposes that each Party restrict practices to the equitable and reasonable use of water while making sufficient data available to the other Parties to verify equitable and reasonable use.

The model is partly based on the structure and organization of the *Delaware River Basin Compact* (DRBC), Pub. L. 87-328, 75 Stat. 688 (1961), and its amendments. This Compact has been extraordinarily successful in resolving interstate conflicts over the water rights and water management of the Delaware River Basin between and among the States of New York, Pennsylvania, New Jersey, and Delaware. Based on the success of the agreement, a companion agreement was made for the Susquehanna River, the *Susquehanna River Basin Compact*, Pub. L. No. 91-575, 84 Stat. 1509 (1970).

Certain provisions of the DRBC are, however, inappropriate for inclusion in the water sharing agreements formulated for other water basins. The reasons for their inapplicability may be the result of differing geophysical or climatic/meteorological differences, differing geopolitical systems or differences in purposes and scopes of the agreement. Therefore, significant modification and revision to the DRBC have been made in this Model Agreement order to to make the model agreement sufficiently flexible for use on both an interstate and international scale and in a variety of geopolitical settings. Certain sections that are Basin-specific have been eliminated. These modifications and revisions resulted from an evaluation of a number of successful water sharing agreements as well as manuscripts, treatises and other studies published by experts and learned scholars.

International agreements provided significant assistance in developing policies and provisions that are science-based and which reflect the globalization of water policy planning. From a policy standpoint, the Model Comprehensive Water Management Agreement adheres to the United Nations *Convention on the Law of the Non-Navigational Uses of International Watercourses*, United Nations Document A/51/869 (1998), signed (but not yet ratified) by the United States as one of the 104 signatories. This Convention attempts to set standards for all international agreements involving the shared use of transboundary water resources. It specifically establishes the dual criteria of "equitable and reasonable utilization" of the water resources and the need to "exchange data and consult on the possible effects of planned measures on the condition" of the water resource. This agreement, as well as other international agreements, is relevant to interstate agreements because of what it offers as guiding principles for the peaceful resolution of water disputes. Much international study has been concentrated on developing successful water sharing agreements and the states with the United States can profit from this study.

COMPREHENSIVE AGREEMENT
CONCERNING THE
SHARED USE OF TRANSBOUNDARY WATER RESOURCES
OF THE
_____ RIVER BASIN

ARTICLE 1C

DECLARATION OF POLICIES AND PURPOSES

§1C-1-01   GENERAL   POLICIES

(a) The water resources of the _____ River Basin have a local, regional and national significance and their comprehensive management for purposes of conservation, utilization, development and control, under appropriate arrangements for intergovernmental cooperation, are public purposes for the respective signatory Parties.

(b) The purposes of this Agreement are to promote interstate [international] comity, to remove causes of present and future controversy, and to efficiently and effectively utilize the shared water resources through integrated management and control.

(c) The general policies of this Agreement include the equitable and reasonable use of the water resources of the _____ River Basin, the need to exchange data and the intent to engage in consultation on the possible effects of planned measures on the condition of the water resource.

(d) The geophysical, climatic/meteorological and other conditions peculiar to the _____ River Basin constitute the basis for this Agreement and its provisions are applicable only to the Basin.

Commentary. Economic growth and prosperity require adequate supplies of quality water on a regular and sustained basis. This means water use must be well managed to ensure the planning, conservation, utilization, development, management and control of water resources on a water basin basis, in a manner that extends beyond political boundaries and encompasses relinquishment of a degree of political sovereignty on the part of the basin States, has been recognized as the only viable way to achieve this objective. Planning at the river basin scale is a recognized basis for adequate water management, especially with regard to access to potable water. *See*, e.g., Tomasz Zylicz, *Environmental Policy Reform in Poland*, Economic Policies for Sustainable Development, Thomas Sterner, ed. (1994). A key challenge for the parties is to make more efficient and productive use of water and to reshape the institutions managing water allocation to better adapt to the increase in water issues, whether they be concerned with water quality or quantity. *See* S. Postel, Dividing the Waters: Food Security, Ecosystem Health, and the New Politics of Scarcity (1996).

One criterion of the *Convention on the Law of the Non-Navigational Uses of International Watercourses*, United Nations Document A/51/869 (1998), makes the definition of the waters to which an agreement applies a mandatory provision in the agreement. The Parties must carefully frame the extent of the specific water resources involved in the agreement. They should identify the type and geographical extent of the water resources to be subject to the Agreement. In order to be formulate an effective agreement, the Parties should analyze the factors which influence the water resource in question, such as the climatology, physiology, geology and the interaction between underground and surface water resources. The analysis must identify pollution sources and the resulting impact on water quality. The purposes and scope of the Agreement will weigh heavily on this assessment, as it will delineate the Parties to be included. The geographic scope of the water resources to be covered by the Agreement should be sufficiently expansive to fully address all the water sharing issues involved. The use of the term "River Basin," if objectionable for any reason by one or more Parties, may be changed to "region," "area," or "system" or any word that accurately describes and encompasses the entirety of the water resources subject to the Agreement

It is important to acknowledge that the Agreement reflects the particular circumstances and compromises reached in its formulation, as applied to the particular basin so that it is recognized the provisions of the agreement are restricted only to those water resources the Parties intend. The inclusion of §1C-1-01(d) may avoid later claims that other rivers and Basins, or other bodies of water, should be dealt with in a similar manner.

Cross-references: §1C-1-02 (purposes of agreement); §1C-1-03 (objectives of agreement); §1C-1-04 (coordination and cooperation); §1C-1-05 (reservation of federal rights); §1C-1-06 (national security; for use in the international context); §2C-1-01 (activation of agreement); §2C-1-02 (consent to jurisdiction)]; §2C-1-03 (duration of agreement); §2C-1-04 (amendments and supplements); §2C-1-05 (powers of sovereign parties; withdrawal); §2C-1-06 (existing agencies); §2C-1-07 (limited applicability); §2C-1-08 (annexes); §2C-2-11 (waters of the basin); §3C-1-01 (commission created); §3C-1-02 (jurisdiction of the commission); §3C-1-03 (commissioners); §3C-1-04 (status, immunities and privileges); §3C-1-05 (commission organization and staffing); §3C-1-06 (rules of procedures); §3C-1-07 (commission administration); §3C-2-01 (general powers and duties); §3C-2-02 (powers and duties reserved to the commissioners); §3C-2-03 (obligations of the commission); §3C-2-04 (regulations; enforcement); §3C-2-05 (prohibited activities); §3C-2-06 (referral and review); §3C-2-07 (advisory committees); §3C-2-08 (reports); §3C-2-09 (condemnation proceedings); §3C-2-10 (meetings, hearings and records); §3C-2-11 (tort liability); §3C-3-01 (coordination and cooperation); §3C-3-02 (project costs and evaluation standards); §3C-3-03 (projects of the signatory parties); §3C-3-04 (cooperative services); §4C-1-01 (joint exercise of sovereignty); §4C-1-02 (interrelationship of water resources); §4C-1-03 (comprehensive water management plan); §4C-1-04 (purpose and objectives of comprehensive water management plan); §4C-1-05 (conditions of comprehensive water management Plan); §4C-1-06 (deviation from comprehensive water management plan); §4C-1-07 (allocation during flood conditions); §4C-1-08 (Allocation during drought conditions); §4C-1-09 (non-impairment of comprehensive water management plan by state action); §4C-2-01 (water allocation, generally); §4C-2-02 (waters not subject to allocation); §4C-2-03 (basin water sources); §4C-2-04 (basin water demands and needs); §4C-2-05 (allocation to equitable and reasonable uses); §4C-2-06 (watershed management); §4C-3-01 (existing rights recognized); §4C-3-02 (flood protection works); §4C-3-03 (minimum flows); §4C-3-04 (withdrawals and diversion; protected areass); §4C-3-05 (water levels protected); §4C-3-06 (augmentation of supply); §4C-3-07, Water Quality; §4C-3-08 (underground water; limit on withdrawals); §4C-3-09 (atmospheric water); §4C-3-10 (recreation); Article 5C (financing); §6C-1-01 (good faith implementation); §6C-1-02 (modification of agreement); §6C-1-03 (material breach); §6C-1-04 (alternative dispute resolution); §6C-1-05 (negotiations and consultations); §6C-1-06 (consultations and mediation); §6C-1-07 (arbitration); §6C-1-08 (right to litigate).

Similar Agreements: *Delaware River Basin Compact*, Pub. L. 87-328, 75 Stat. 688 (1961); *Susquehanna River Basin Compact*, Pub. L. No. 91-575, 84 Stat. 1509 (1970); *Convention on the Law of the Non-Navigational Uses of International Watercourses*, United Nations Document A/51/869 (1998).

§1C-1-02 PURPOSES AND SCOPE OF AGREEMENT

By this Agreement, the signatory Parties agree to the following:

(a) to provide for the most efficient use of the Waters of the Basin through comprehensive, basin-wide management;

(b) to provide for the equitable and reasonable utilization of the water resources, including surface, underground, and atmospheric supplies;

(c) to provide for the maintenance of the quality of water within the Basin;

(d) to remove all causes, present and future, which might lead to controversies or conflicts; to promote interstate [international] comity; and

(e) to minimize economic and social disruption and dislocation through implementation of drought strategies and flood control programs.

(f) to limit this Agreement to the water resources arising within the drainage basin of the _____ River; and

(g) the Parties to this Agreement are the States of [Party 1] and [Party 2].

Commentary. The purpose statement is intended to be indefinite in order to allow flexibility in the management of the shared water resource. While the Parties may wish to be more specific when using this model Agreement, the need for such specificity may mean the agreement presented at Model B is more appropriate.

Goldfarb has noted that "integration of water quality and quantity programs is one aspect of, and is inseparable from, comprehensive, coordinated, multipurpose, basinwide water and related resource planning and management." *See* William Goldfarb, *The Trend Toward Judicial Integration of Water Quality and Quantity Management: Facing the New Century* in Water Resources Administration in the United States (M. Reuss, ed., 1993).

Many recent international agreements, most notably the *Convention on the Law of the Non-Navigational Uses of International Watercourses*, United Nations Document A/51/869 (1998), establish as a principle purpose of all transboundary water sharing agreements the "equitable and reasonable utilization" of water.

If the agreement is to control the use of underground and atmospheric water as well as surface water, the intent to encompass underground and atmospheric water within the scope of the agreement should be made clear at this point. Failure to specifically deal with the issue of underground water withdrawal has led to litigation in the context of existing interstate compacts in the United States. It is therefore recommended that underground and atmospheric water be included within the scope of the agreement. As technology advances, use of and control of atmospheric water may also become more commonplace, and co ideration should be given to dealing with the potential for such use and control before it becomes established and results in unexpected interference with the provisions made for underground and surface water.

This provision further defines the geographical description of the international watercourse in question, as required by the *Convention on the Law of the Non-Navigational Uses of International Watercourses*, United Nations Document A/51/869 (1998).

Cross-references: §1C-1-01 (general policies); §1C-1-03 (objectives of agreement); §1C-1-04 (coordination and cooperation); §1C-1-05 (reservation of federal rights); §1C-1-06 (national security; for use in the international context); §2C-1-01 (activation of agreement); §2C-1-02 (consent to jurisdiction); §2C-1-03 (duration of agreement); §2C-1-04 (amendments and supplements); §2C-1-05 (powers of sovereign parties; withdrawal); §2C-1-06 (existing agencies); §2C-1-07 (limited applicability); §2C-1-08 (annexes); §2C-2-06 (equitable and reasonable apportionment); §2C-2-07 (flood); §2C-2-09 (party or parties); §2C-2-10 (underground water); §2C-2-11 (waters of the basin); §3C-1-01 (commission created); §3C-1-02 (jurisdiction of the commission); §3C-1-03 (commissioners); §3C-1-04 (status, immunities and privileges); §3C-1-05 (commission organization and staffing); §3C-1-06 (rules of procedures); §3C-1-07 (commission administration); §3C-2-01 (general powers and duties); §3C-2-02 (powers and duties reserved to the commissioners); §3C-2-03 (obligations of the commission); §3C-2-04 (regulations; enforcement); §3C-2-05 (prohibited activities); §3C-2-06 (referral and review); §3C-2-07 (advisory committees); §3C-2-08 (reports); §3C-2-09 (condemnation proceedings); §3C-2-10 (meetings, hearings and records); §3C-2-11 (tort liability); §3C-3-01 (coordination and cooperation); §3C-3-02 (project costs and evaluation standards); §3C-3-03 (projects of the signatory parties); §3C-3-04 (cooperative services); §4C-1-01 (joint exercise of sovereignty); §4C-1-02 (interrelationship of water resources); §4C-1-03 (comprehensive water management plan); §4C-1-04 (purpose and objectives of comprehensive water management plan); §4C-1-05 (conditions of comprehensive water management Plan); §4C-1-06 (deviation from comprehensive water management plan); §4C-1-07 (allocation during flood conditions); §4C-1-08 (Allocation during drought conditions); §4C-1-09 (non-impairment of comprehensive water management plan by state action); §4C-2-01 (water allocation, generally); §4C-2-02 (waters not subject to allocation); §4C-2-03 (basin water sources); §4C-2-04 (basin water demands and needs); §4C-2-05 (allocation to equitable and reasonable uses); §4C-2-06 (watershed management); §4C-3-01 (existing rights recognized); §4C-3-02 (flood protection works); §4C-3-03 (minimum flows); §4C-3-04 (withdrawals and diversion; protected areass); §4C-3-05 (water levels protected); §4C-3-06 (augmentation of supply); §4C-3-07 (water quality); §4C-3-08 (underground water; limit on withdrawals); §4C-3-09 (atmospheric water); §4C-3-10 (recreation); Article 5C (financing); §6C-1-01 (good faith implementation); §6C-1-02 (modification of agreement); §6C-1-03 (material breach); §6C-1-04 (alternative dispute resolution); §6C-1-05 (negotiations and consultations); §6C-1-06 (consultations and mediation); §6C-1-07 (arbitration); §6C-1-08 (right to litigate).

Similar Agreements: *Klamath River Basin Compact*, 71 Stat. 497 (1957); *Delaware River Basin Compact* (DRBC), Pub. L. 87-328, 75 Stat. 688 (1961); *Susquehanna River Basin Compact*, Pub. L. No. 91-575, 84 Stat. 1509 (1970); *Alabama-Coosa-Tallapoosa River Basin Compact*, O.C.G.A. 12-10-110 (1997); *The Appalachicola-Chatahoochee-Flint River Basin Compact* (1997); *Convention on Wetlands of International Importance, Especially as Waterfall Habitat,* (1971); *Convention on Biological Diversity*, UNEP/Bio.Div/N7-INC.5/4 (1992); *The Rio Declaration on the Environment and Development*, 31 I.L.M. 874 (1992); *The Declaration of Principles on Interim Self-Government Arrangements*, Sep. 13, 1993, Israel.-PLO, 32 I.L.M. 1525 (1993); *Convention on the Law of the Non-Navigational Uses of International Watercourses*, United Nations Document A/51/869 (1998);

§1C-1-03 OBJECTIVES OF AGREEMENT (Optional)

The Parties agree to the following objectives:

(a) To cooperate in all fields of sustainable development, utilization, management and conservation of the water and related resources of the waters of the _____ River Basin including, but not limited to agricultural irrigation, domestic and commercial use, industrial and mining use, hydropower, navigation, flood control, fisheries, recreation and tourism, and maintenance of natural water environments in a manner to optimize the multiple use and mutual benefits of all riparians and to minimize the harmful effects that might result from natural occurrences and man-made activities.

(b) To promote, support, cooperate and coordinate in the development of the full potential of sustainable waters of the _____ River Basin, with emphasis and preference on joint and/or basin-wide development projects and basin programs through the formulation of a basin development plan that would be used to identify, categorize and prioritize the projects and programs to seek assistance for and to implement at the basin level.

MODEL WATER SHARING AGREEMENTS 91

    (c) To protect the environment, natural resources, aquatic life and conditions, and ecological balance of the waters of the _____ River Basin from pollution or other harmful effects resulting from any development plans and uses of water and related resources in the Basin.

    (d) To cooperate on the basis of sovereign equality and territorial integrity in the utilization and protection of water resources of the waters of the _____ River Basin.

    (e) To utilize the waters of the waters of the _____ River Basin in a reasonable and equitable manner in their respective territories.

Commentary: This provision provides a framework for the Parties in the development of their individual water development planning. It recognizes that there are certain fundamental principles that each Party must follow in their rationale management of water resources. It would be irrational for one Party to agree to "equitable and reasonable utilization" when it does not follow a similar philosophy within its own borders. These general objectives and principles improve the attainment of purposes of the water sharing.

Cross-references: §1C-1-01 (general policies); §1C-1-02 (purposes of agreement); §1C-1-04 (coordination and cooperation); §1C-1-05 (reservation of federal rights); §1C-1-06 (national security; for use in the international context); §2C-2-07 (flood); §2C-2-09 (party or parties); §4C-1-02 (interrelationship of water resources); §4C-1-03 (comprehensive water management plan); §4C-1-04 (purpose and objectives of comprehensive water management plan); §4C-1-05 (conditions of comprehensive water management Plan); §4C-1-06 (deviation from comprehensive water management plan); §4C-1-07 (allocation during flood conditions); §4C-1-08 (Allocation during drought conditions); §4C-2-01 (water allocation, generally); §4C-2-02 (waters not subject to allocation); §4C-2-05 (allocation to equitable and reasonable uses); §4C-2-06 (watershed management); §4C-3-01 (existing rights recognized); §4C-3-02 (flood protection works); §4C-3-03 (minimum flows); §4C-3-04 (withdrawals and diversion; protected areass); §4C-3-05 (water levels protected); §4C-3-06 (augmentation of supply); §4C-3-07 (water quality); §4C-3-08 (underground water; limit on withdrawals); §4C-3-09 (atmospheric water); §4C-3-10 (recreation).

Similar Agreements: *Convention Between Switzerland and Italy Concerning the Protection of Italo-Swiss Waters Against Pollution*, UNTS, Vol. 957, 277 (1972); *The North American Agreement on Environmental Cooperation between the Government of the United States of America, the Government of Canada, and the Government of the United Mexican States*, 32 I.L.M. 1480 (1993); *Agreement on the Cooperation for the Sustainable Development of the Mekong River Basin*, 34 ILM 864 (1995).

§1C-1-04   COORDINATION AND COOPERATION

    (a) Each of the Parties pledges to support implementation of the provisions of this Agreement, and covenants that its officers and agencies will not hinder, impair,

or prevent any other Party carrying out any provision or recommendation of this Agreement.

(b) The Parties shall at all times endeavor to agree on the interpretation and application of this Agreement, and shall make every attempt through cooperation and consultations to arrive at a mutually satisfactory resolution of any matter that might affect its operation.

(c) The Parties acknowledge that their respective governmental organizations shall provide the information necessary to assist in the equitable and reasonable utilization of the waters of the Basin. Such information shall include, but not be limited to, all planning and management activities and water projects affecting their common boundary water resources.

(d) The Parties further acknowledge that all states are expected to conduct themselves with an absence of malice and with no intention to seek unconscionable advantage, or otherwise be deceitful.

Commentary. A Party normally enters into any international agreement with a position of self-interest. In the negotiations, each Party seeks the rights and authorities critical to certain political, economic or social objectives while ceding less critical rights and authorities to the other nations. While accepting this fact, all Parties have a duty to cooperate and negotiate in good faith. This principle is the foundation of international law, and it applies in all relations between sovereign states.

Cross-references: §1C-1-01 (general policies); §1C-1-02 (purposes of agreement); §1C-1-06 (national security); §2C-1-01 (activation of agreement); §2C-1-03 (duration of agreement); §2C-1-04 (amendments and supplements); §2C-1-05 (powers of sovereign parties; withdrawal); §2C-1-06 (existing agencies); §2C-1-07 (limited applicability); §2C-1-08 (annexes); §2C-2-09 (party or parties); §3C-1-02 (jurisdiction of the commission); §3C-1-03 (commissioners); §3C-1-04 (status, immunities and privileges); §3C-1-05 (commission organization and staffing); §3C-1-06 (rules of procedures); §3C-1-07 (commission administration); §3C-2-01 (general powers and duties); §3C-2-02 (powers and duties reserved to the commissioners); §3C-2-03 (obligations of the commission); §3C-2-04 (regulations; enforcement); §3C-2-05 (prohibited activities); §3C-2-06 (referral and review); §3C-2-07 (advisory committees); §3C-2-08 (reports); §3C-2-09 (condemnation proceedings); §3C-2-10 (meetings, hearings and records); §3C-2-11 (tort liability); §3C-3-01 (coordination and cooperation); §3C-3-02 (project costs and evaluation standards); §3C-3-03 (projects of the signatory parties); §3C-3-04 (cooperative services); §4C-1-01 (joint exercise of sovereignty); §4C-1-02 (interrelationship of water resources); §4C-1-03 (comprehensive water management plan); §4C-1-04 (purpose and objectives of comprehensive water management plan); §4C-1-05 (conditions of comprehensive water management Plan); §4C-1-06 (deviation from comprehensive water management plan); §4C-1-07 (allocation during flood conditions); §4C-1-08 (Allocation during drought conditions); §4C-1-09 (non-impairment of comprehensive water management plan by state action); §4C-2-01 (water allocation, generally); §4C-2-02 (waters not subject to allocation); §4C-2-05 (allocation to equitable and reasonable uses); §4C-2-06 (watershed management); §4C-3-01 (existing rights recognized);

§4C-3-02 (flood protection works); §4C-3-03 (minimum flows); §4C-3-04 (withdrawals and diversion; protected areass); §4C-3-05 (water levels protected); §4C-3-06 (augmentation of supply); §4C-3-07 (water quality); §4C-3-08 (underground water; limit on withdrawals); §4C-3-09 (atmospheric water); §4C-3-10 (recreation); Article 5C (financing); §6C-1-01 (good faith implementation); §6C-1-02 (modification of agreement); §6C-1-03 (material breach); §6C-1-04 (alternative dispute resolution); §6C-1-05 (negotiations and consultations); §6C-1-06 (consultations and mediation); §6C-1-07 (arbitration); §6C-1-08 (right to litigate).

Similar Agreements: Agreement Between the People's Republic of Bulgaria and the Republic of Turkey Concerning Co-operation in the Use of the Waters of Rivers Flowing through the Territory of Both Countries, UNTS, Vol. 807, 117 (1968); *Stockholm Declaration of the United Nations Conference on the Human Environment*, 11 I.L.M. 1416 (1972); *Treaty for Amazonian Cooperation*, 17 ILM 1046 (1978); Convention on Long-Range Transboundary Air Pollution, 181 I.L.M. 1442 (1979); Vienna Convention for the Protection of the Ozone Layer (1985); Convention Between the Federal Republic of Germany and the Czech and Slovak Federal Republic and the European Economic Community on the International Commission for the Protection of the Elbe, International Environmental Law, Multilateral Agreements, 976:90/1 (1990); *Convention on the Protection and Use of Transboundary Watercourses and International Lakes*, 31 I.L.M. 1312 (1992); *The North American Agreement on Environmental Cooperation between the Government of the United States of America, the Government of Canada, and the Government of the United Mexican States*, 32 I.L.M. 1480 (1993); *Convention on the Law of the Non-Navigational Uses of International Watercourses*, United Nations Document A/51/869 (1998); *Treaty of Peace between the State of Israel and the Hashemite Kingdom of Jordan*, 34 I.L.M. 43 (1994).

§1C-1-05 PRESERVATION OF FEDERAL RIGHTS (For U.S. use only)

Nothing in this agreement shall be deemed:

(a) To impair or affect any rights or powers of the United States, its agencies or instrumentalities, in and to the use of the waters of the _____ River Basin nor its capacity to acquire rights in and to the use of said waters;

(b) To subject any property of the United States, its agencies, or instrumentalities to taxation by either Party, nor to create an obligation on the part of the United States, its agencies, or instrumentalities, by reason of the acquisition, construction or operation of any property or works of whatsoever kind, to make any payments to any State or political subdivision thereof, State agency municipality, or entity whatsoever in reimbursement for the loss of taxes;

(c) To subject any property of the United States, its agencies, or instrumentalities, to the laws of any Party to an extent other than the extent to which these laws would apply without regard to the agreement.

Commentary: These sections may be included in agreements between states of the United States. They are probably unnecessary to preserve federal rights, but inasmuch as Congress must approve the agreement, the inclusion of these provisions may make it easier to obtain that approval.

Cross References: §1C-1-01 (general policies); §1C-1-02 (purposes of agreement); §1C-1-03 (objectives of agreement); §2C-1-03 (duration of agreement); §2C-1-04 (amendments and supplements); §2C-1-05 (powers of sovereign parties; withdrawal); §4C-1-01 (joint exercise of sovereignty).

Similar Agreements: *Republican River Compact*, 57 Stat. 86 (1943); *Belle Fourche River Compact*, 58 Stat. 94 (1944); *Pecos River Compact*, 63 Stat. 159 (1948); *Snake River Compact*, 64 Stat. 29 (1949); *Yellowstone River Compact,* 65 Stat. 663 (1950); *Canadian River Compact,* 66 Stat. 74, (1952); *Klamath River Basin Compact*, 71 Stat. 497 (1957); *Bear River Compact*, 72 Stat. 38 (1955), amended 94 Stat. 4, Art. XIII (2) (1980); *Upper Colorado River Basin Compact*, 63 Stat. 31 (1949).

§1C-1-06 NATIONAL SECURITY (For international use)

(a) Nothing in this Agreement shall be construed to require any Party to make available or provide access to information the disclosure of which it determines to be contrary to its essential security interests.

(b) Nothing in this Agreement shall be construed to prevent any Party from taking any actions that it consider necessary for the protection of its essential security interests relating to a formal declaration of war.

Commentary: National security concerns will necessarily take precedence over any program of water management and the exchange of data.

Cross-references: §1C-1-01 (general policies); §1C-1-02 (purposes of agreement); §1C-1-03 (objectives of agreement); §1C-1-04 (coordination and cooperation); §2C-1-04 (amendments and supplements); §2C-1-05 (powers of sovereign parties; withdrawal); §2C-2-09 (party or parties).

Similar Agreements: *The North American Agreement on Environmental Cooperation between the Government of the United States of America, the Government of Canada, and the Government of the United Mexican States*, 32 I.L.M. 1480 (1993).

ARTICLE 2C

GENERAL PROVISIONS

Part 1 General Obligations

§2C-1-01   EFFECTIVE DATE

Alternative 1: (For international use)

This Agreement shall become operative when approved by the appropriate governing authorities of all Parties. The agreement will go into full force and effect at 12:01 a.m. [time zone] on the day immediately following the final act necessary for approval of the agreement, as defined by the domestic law of each Party, by the last Party to give such approval.

Alternative 2: (For U.S. use)

This Agreement shall become operative when, subnsequent to adoption of each of the States, the Congress of the United States adopts legislation providing, among other things, that:

(a) Any equitable and reasonable utilization hereafter made by the United States, or those acting by or under its authority, within a State, of the waters allocated by this agreement, shall be within the allocations hereinabove made for use in that State and shall be taken into account in determining the extent of use within that State.

(b) The United States, or those acting by or under its authority, in the exercise of rights or powers arising from whatever jurisdiction the United States has in, over and to the waters of the _____ River and all its tributaries, shall recognize, to the extent consistent with the best utilization of the waters for multiple purposes, that equitable and reasonable use of the waters within the Basin is of paramount importance to development of the Basin, and no exercise of such power or right by the United States government or those acting under its authority that would interfere with the full equitable and reasonable use of the waters shall be made except upon a determination, giving due consideration to the objectives of this agreement and after consultation with all interested Federal agencies and the State officials charged with the administration of this agreement, that such exercise is in the interest of the best utilization of such waters for multiple purposes.

(c) The United States or those acting by or under its authority will recognize any established use, for domestic and irrigation purposes, of the apportioned waters which may be impaired by the exercise of Federal jurisdiction in, over, and to such waters; provided, that such use is being exercised equitably and reasonably, is valid under the laws of the appropriate state and in conformity with this agreement at the time of the impairment thereof and was validly initiated under state law prior to the initiation or authorization of the Federal program or project which causes such impairment.

Commentary: Any agreement of this nature should specify the date or conditions upon which it will take effect. In the case of agreements between states of the United States, the conditions with respect to Congress are designed to provide some measure of protection against subsequent federal action that might disturb the allocation system agreed upon by the contracting Parties. Despite the requirement of federal approval of interstate compacts, the federal government is not normally a Party to those agreements and might not be bound by the provisions of those agreements unless there is specific legislation committing the federal government to be so bound. The provisions of Alternative 2, modeled after the *Republican River Compact*, 57 Stat. 86 (1943) and the *Bell Fourche Compact*, 58 Stat. 94 (1944) condition the effectiveness of the agreement on passage of such legislation by Congress and also establish a basis for compensation for takings under the Fifth Amendment should a subsequent Congress decide to take action contrary to that commitment (a later Congress has the power to set aside the actions of an earlier Congress, but the question of takings and just compensation then arises.) If these conditions are not incorporated, the States making the agreement may later find that federal actions render their agreement ineffective.

Cross-references: §2C-1-03 (duration of agreement); §2C-1-04 (amendments and supplements); §2C-1-05 (powers of sovereign parties; withdrawal); §2C-2-09 (party or parties); §6C-1-02 (modification of agreement).

Similar Agreements: *Republican River Compact*, 57 Stat. 86 (1943); *Bell Fourche Compact*, 58 Stat. 94 (1944); *Delaware River Basin Compact*, Pub. L. 87-328, 75 Stat. 688 (1961); *Susquehanna River Basin Compact*, Pub. L. No. 91-575, 84 Stat. 1509 (1970); *Convention on the Protection and Use of Transboundary Watercourses and International Lakes*, 31 I.L.M. 1312 (1992); *Convention on the Law of the Non-Navigational Uses of International Watercourses*, United Nations Document A/51/869 (1998).

§2C-1-02 CONSENT TO JURISDICTION (for U.S. use)

This agreement shall be effective upon the United States Congress giving its consent for the United States to be named and joined as a Party defendant or otherwise in any case or controversy involving the construction or application of this agreement in which one or more of the States is a plaintiff, without regard to any requirement as to the sum or value in controversy or diversity of citizenship of Parties to the case or controversy.

Commentary: The predominance of federal interests in water resources makes it likely that any litigation concerning the agreement between States will involve federal interests. The doctrine of sovereign immunity could prevent joinder of the federal interests as Parties to the suit absent a waiver of sovereign immunity. The discretionary devision not to join federal parties led to dismissal of a suit filed by Texas against New Mexico in 1951 to enforce certain provisions of the *Rio Grande Compact of 1938*, 53 Stat. 785, 938. The Supreme Court dismissed the case because the federal government was not joined as a party, but had important interests that would be affected by any such suit. *Texas v. New Mexico*, 352 U.S. 991, 957. In 1952, Congress enacted the *McCarren Amendment,* 43 U.S.C. 66, which waived federal sovereign immunity to be joined in general stream adjudications. However, in most cases, the agreement will cover management and development issues that reach beyond general stream adjudication. The Parties should consider including such a waiver of sovereign immunity as a condition to effectiveness of the agreement. They may also wish to add a provision granting jurisdiction over any such cases to the District Courts, which may be preferable to the Supreme Court as the initial forum for resolving certain types of disputes. The *Red River Compact*, 94 Stat. 3305 (1980) takes this approach.

Cross References: §1A-1-04 (preservation of federal right); §1A-1-05 (national security); §2C-2-09 (party or parties).

Similar Agreements: *Kansas-Nebraska Big Blue River Compact*, 86 Stat. 193 (1972); *Red River Compact*, 94 Stat. 3305 (1980).

§2C-1-03 DURATION OF AGREEMENT (Optional)

(a) **The Parties intend that the duration of this Agreement shall be for an initial period of [___] years from its effective date.**

(b) **In the event that this Agreement should be terminated by operation of paragraph (a) above, the management structure for the Agreement shall be dissolved, its assets and liabilities transferred, and its corporate affairs wound up, in such manner as may be provided by agreement of the signatory Parties.**

Commentary: The Parties may prefer to establish no duration and rely on later provisions to modify or terminate the agreement. However, two significant principles are established by this provision. First, setting a duration for an extended period of time, recommended at 50 years, allows for predictability in terms of water resources development; it also allows sufficient time to recover capital costs in the financing of projects. Second, establishing a duration ensures that the Parties reconsider the Agreement only after a sufficient hydrologic record is established. It should be noted, however, that this provision greatly impacts on the exercise of sovereignty of the Parties involved.

Cross-references: §2C-1-01 (activation of agreement); §2C-1-02 (consent to jurisdiction); §2C-1-04 (amendments and supplements); §2C-1-05 (powers of sovereign parties; withdrawal); §2C-2-09 (party or parties); §6C-1-02 (modification of agreement).

Similar Agreements: *Delaware River Basin Compact*, Pub. L. 87-328, 75 Stat. 688 (1961); *Susquehanna River Basin Compact*, Pub. L. No. 91-575, 84 Stat. 1509 (1970).

§2C-1-04 POWERS OF SOVEREIGN PARTIES; WITHDRAWAL
(Optional, international use only)

(a) Nothing in this Agreement shall be construed to relinquish the functions, powers or duties of the government of any signatory Party with respect to the control of any navigable waters within the Basin, nor shall any provision hereof be construed in derogation of any of the powers of the Parties to regulate commerce within their sovereign borders. The power and right of any sovereign Party to withdraw from this Agreement or to revise or modify the terms, conditions and provisions under which it may remain a Party is recognized by the signatory Parties. Notification of the withdrawal must be made (___) months in advance of the prospective withdrawal.

(b) Inasmuch as the other Parties may have committed resources to the comprehensive joint management of the waters of the basin, the withdrawing Party shall equitably compensate the other Parties who have relied on the term of the agreement to justify their capital costs for facilities and other appurtenances.

Commentary. This provision acknowledges the inherent sovereignty of the individual parties and recognizes that any relinquishment of sovereignty to limited solely to the purposes of this Agreement. It does, however, recognize that some Parties may have committed financial and other resources to capital projects and that the early withdrawal of one Party will not allow them to recover those costs. Consequently, the part (b) of this provision provides for equitable compensation for the unrealized recovery of capital expenditures.

Cross-references: §2C-2-09 (party or parties); §2C-2-11 (waters of the basin); §6C-1-01 (good faith implementation); §6C-1-02 (modification of agreement).

Similar Agreements: *Delaware River Basin Compact*, Pub. L. 87-328, 75 Stat. 688 (1961); *Susquehanna River Basin Compact*, Pub. L. No. 91-575, 84 Stat. 1509 (1970); *Apalachicola-Chattahoochee-Flint River Basin Compact*, O.C.G.A. 12-10-100 (1997); *Alabama-Coosa-Tallapoosa River Basin Compact*, O.C.G.A. 12-10-110 (1997); *Convention on the Protection and Use of Transboundary Watercourses and International Lakes*, 31 I.L.M. 1312 (1992); *The North American Agreement on Environmental Cooperation between the Government of the United States of America, the Government of Canada, and the Government of the United Mexican States*, 32 I.L.M. 1480 (1993); *Agreement on the Cooperation for the Sustainable Development of the Mekong River Basin*, 34 ILM 864 (1995).

§2C-1-05 AMENDMENTS AND SUPPLEMENTS (Optional)

The provisions of this agreement shall remain in full force and effect until amended by action of the governing bodies of the Parties and consented to and approved by any other necessary authority in the same manner as this agreement is required to be ratified to become effective.

Commentary. Agreements may, over time, cease to operate as well as initially hoped. Some amendment process should be specified. In some cases, the approval of another institution may be required. If, for example, the agreement is between states of the United States, the United States Constitution probably requires Congressional approval of any amendment as well as approval of the original agreement, unless the agreement provides for a different method of amendment. In this latter case, The Congressional approval of the initial agreement would implicitly grant consent to modify the agreement in accordance with the terms of the agreement. If the agreement is between sovereign nations, the references to other "necessary authority" would probably be omitted, but the particular circumstances of each case must be considered.

Cross-references: §2C-1-07 (limited applicability); §2C-2-09 (party or parties); §6C-1-02 (modification of agreement).

Similar Agreements: *Delaware River Basin Compact*, Pub. L. 87-328, 75 Stat. 688 (1961); *Susquehanna River Basin Compact*, Pub. L. No. 91-575, 84 Stat. 1509 (1970); *Convention on the Protection and Use of Transboundary Watercourses and International Lakes*, 31 I.L.M. 1312 (1992); Agreement *on the Cooperation for the Sustainable Development of the Mekong River Basin*, 34 ILM 864 (1995).

§2C-1-06 EXISTING AGENCIES

It is the purpose of the signatory Parties to preserve and utilize the functions, powers and duties of existing offices and agencies of government to the extent not inconsistent with the Agreement, and the institution established to enforce this Agreement is authorized and directed to utilize and employ such offices and agencies for the purpose of this Agreement to the fullest extent it finds feasible and advantageous.

Commentary: The use of existing offices and agencies prevents duplication, and consequently costs, of effort in the data collection and management of the water resource subject to the Agreement.

Cross-references: §2C-2-09 (party or parties); §3C-1-05 (commission organization and staffing); §3C-1-06 (rules of procedures); §3C-1-07 (commission administration); §3C-3-04 (cooperative services).

Similar Agreements: *Delaware River Basin Compact,* Pub. L. 87-328, 75 Stat. 688 (1961); *Susquehanna River Basin Compact,* Pub. L. No. 91-575, 84 Stat. 1509 (1970); Convention on the Use of Transboundary Watercourses and International Lakes, 31 I.L.M. 1312 (1992).

§2A-1-07 SEVERABILITY

Should a tribunal of competent jurisdiction hold any part of this Agreement to be void or unenforceable, it shall be considered severable from those portions of the greement capable of continued implementation in the absence of the voided provisions. All other severable provision capable of continued implementation shall continue in full force and effect.

Commentary: The drafters of the Agreement should consider whether they wish this clause to be included. The advantage of such a clause is that it avoids the possibility of having the entire Agreement become null and void if any part is found to be void or unenforceable.

Cross-references: §2A-1-04 (amendments and supplements).

Similar Agreements: *Yellowstone River Compact,* 65 Stat. 663 (1950); *Sabine River Compact,* 68 Stat. 690 (1953); *Klamath River Basin Compact,* 71 Stat. 497 (1957); *Delaware River Basin Compact,* Pub. L. 87-328, 75 Stat. 688 (1961); *Susquehanna River Basin Compact,* Pub. L. No. 91-575, 84 Stat. 1509 (1970).

§2C-1-08 ANNEXES

The Annexes to this Agreement to the extent consistent with the objectives and intent of the Agreement constitute an integral part of the Agreement.

Commentary: An effective water management agreement will necessarily contain detailed information and data of a procedural nature. While such information may be essential to the effectiveness of the particular agreement, its inclusion in the main body of the agreement may take away from the essence of the contractual nature of the agreement. The use of annexes minimizes this effect.

Cross-references: Art. 6C (financing); §6C-1-07 (arbitration).

Similar Agreements: ASEAN Agreement on the Conservation of Nature and Natural Resources, (1985); *The North American Agreement on Environmental Cooperation between the Government of the United States of America, the Government of Canada, and the Government of the United Mexican States,* 32 I.L.M. 1480 (1993).

Part 2 Definitions

§2C-2-01 ATMOSPHERIC WATER

The phrases "atmospheric water" means water produced by artificial changes in the composition, motions, and resulting behavior of the atmosphere or clouds, including fog, or with the intent of inducing changes in precipitation by use of electrical device, lasers, alterations of the earth's surface, or cloud seeding.

Commentary: This definition is consistent with the definitions usually used in State and federal laws on weather modification. *See* generally Robert Beck, *Augmenting the Available Water Supply* in 1 Waters and Water Rights § 3.04; Ray Jay Davis, *Four Decades of American Weather Modification Law*, 19 J. Weather Modification 102 (1987).

Cross-references: §4C-3-09 (atmospheric water).

§2C-2-02 _____ BASIN

_____ Basin" means the area of drainage into the _____ River and its tributaries, [and] aquifers underlying the drainage, or only the aquifers themselves.

Commentary: The Agreement could include the total surface area of drainage throughout the Basin and contain aquifers underlying the surface drainage. Some tributaries can be connected to the underlying aquifers holding the underground water. Some of the aquifers could be connected to more than one of the surface water basins. The geographic scope of the Agreement should be defined to ensure there are no future disagreements about what lands are or are not covered by the Agreement. A map may be incorporated, but care should be taken that the map is cartographically accurate. Because the map is likely to be at a scale too small for precise delineation of boundaries, it should be made clear that it is for general reference only. In the event of a dispute over land or within the defined _____ River and its tributaries, the actual limits of the watershed as determined on the ground should be controlling

Cross References: §3C-1-01 (commission created); §3C-1-02 (jurisdiction of the commission); §3C-2-01 (general powers and duties); §3C-2-02 (powers reserved to the commissioners); §3C-2-03 (obligations of the commission); §3C-2-04 (regulations; enforcement); §3C-3-01 (coordination and cooperation); §3C-3-03 (projects of the signatory parties); § 4C-1-01 (joint exercise of sovereignty); §4C-1-03 (comprehensive water management plan); §4C-1-04 (purpose and objectives); §4C-1-05 (conditions of comprehensive plan); §4C-1-07 (allocations during flood conditions); §4C-1-09 (non-impairment of comprehensive water plan by state action); §4C-2-01 (basin water sources); §4C-2-03 (basin water demands and needs); §4C-2-06 (watershed management); §4C-3-01 (existing rights recognized); §4C-3-02 (flood protection works); §4C-3-04 (withdrawals and diversions); § 4C-3-06 (augmentation of supply); §4C-3-07 (water quality).

§2C-2-03   COMPREHENSIVE   WATER   MANAGEMENT   PLAN

"Comprehensive Water Management Plan" means a plan which considers the interrelationship of transboundary water resources, describes current and prospective water uses, identifies water supplies, and matches these supplies to water uses. It also identifies needed water-related management measures, facility needs and costs, addresses environmental concerns, and offers program and policy recommendations to better manage the basin's water resources and water quality. This plan is for the long term to produce the highest quality and most efficient use of water resources for the greatest benefit to the public and the environment.

Commentary: Comprehensive water management plans generally are guides for the orderly development and management of water resources. Generally, these plans span a 50-year horizon, consider population growth, development and availability of new water supplies, water transfers from one basin to another, data sources and methodologies, cost of water, regional or sub-regional plans, regulatory issues, economic development, specific projects, health and public safety issues, and other concerns. The Parties should negotiate the time horizon and the frequency for updating the plan.

Cross-references: §4C-1-01 (Joint Exercise of Sovereignty); §4C-1-02 (Interrelationship of Water Resources); §4C-1-03 (comprehensive water management plan); §4C-1-04 (purpose and objectives of comprehensive Water management plan); §4C-1-05 (conditions of comprehensive water management Plan); §4C-1-06 (deviation from comprehensive water management plan); §4C-1-07 (allocation during flood conditions); §4C-1-08 (Allocation during drought conditions); §4C-1-09 (non-impairment of comprehensive water Management plan by state action); §4C-2-01 (water allocation, generally); §4C-2-03 (basin water sources); §4C-2-04 (Basin Water Demands and Needs); §4C-2-05 (Allocation to Equitable and Reasonable Uses); §4C-2-06 (watershed management); §4C-3-01 (existing rights recognized); §4C-3-02 (flood protection works); §4C-3-03 (Minimum Flows); §4C-3-04 (Withdrawals and Diversion; Protected Areas); §4C-3-05 (Water Levels Protected); §4C-3-06 (Augmentation of Supply); §4C-3-07 (Water Quality); §4C-3-08 (Underground Water; Limit on Withdrawals); §4C-3-09 (Atmospheric Water); §4C-3-10 (Recreation).

§2C-2-04   CONSERVATION MEASURES

"Conservation measures" refers to any measures adopted by a water right holder, or several water right holders acting in concert pursuant to a conservation agreement reviewed and approved by the Commission, as being appropriate water-saving strategies for purposes of the Comprehensive Water Management Plan, to reduce the withdrawals and/or consumptive uses, including, but not limited to:

(a)   Improvements in water transmission and water use efficiency;
(b)   Reduction in water use;
(c)   Enhancement of return flows; and
(d)   Reuse of return flows.

Commentary: Sustainable development requires steps to conserve the waters of the basin. This definition limits the application of the term "conservation measures" to practices that have been reviewed and approved by the Commission as being appropriate water-saving strategies for the purposes of the Comprehensive Water Management Plan. Specifically excluded from this definition are practices applied to native or naturally occurring waters, return flows from other water rights, or other water sources not associated with the water right holder or sought by the applicant.

Nothing in this Model Agreement attempts to detail appropriate conservation measures. Such efforts as improved efficiency in manufacturing processes, the substitution of drip irrigation for sprinklers, or the introduction by a public water supply enterprise of requirement that customers use low flow toilets or showerheads would all be appropriate examples. The Model Agreement leaves the precise details regarding the suitability of these or other possible conservation measures to be developed by the regulatory and planning processes prescribed for the State Agency.

Cross-references: §4C-1-08 Allocation under Drought Conditions.

§2C-2-05   DROUGHT

"Drought" conditions means conditions brought about by the lack of precipitation or water stored in the soil, in a quantity agreeable to the Parties, from the mean annual precipitation and water measured in soil.

Commentary: Management action will arise from a drought, or lack of mean annual rainfall, but could arise from other causes as well, such as the collapse of a dam with the resulting draining of a reservoir on which the Commission users depend. The definition should be determined, in large measure, by the use intended. Then a "drought management strategy" would be a specific course of conduct planned by the Commission as a necessary or appropriate response to the lack of precipitation.

Cross-references: §4C-1-03 (comprehensive water management plan); §4C-1-04 (purpose and objectives of comprehensive Water management plan); §4C-1-05 (conditions of comprehensive water management Plan); §4C-1-06 (deviation from comprehensive water management plan); §4C-1-08 (Allocation during drought conditions); §4C-1-09 (non-impairment of comprehensive water Management plan by state action); §4C-2-01 (water allocation, generally); §4C-2-03 (basin water sources); §4C-2-04 (Basin Water Demands and Needs); §4C-2-05 (Allocation to Equitable and Reasonable Uses); §4C-2-06 (watershed management); §4C-3-01 (existing rights recognized); §4C-3-03 (Minimum Flows); §4C-3-04 (Withdrawals and Diversion; Protected Areas); §4C-3-05 (Water Levels Protected); §4C-3-06 (Augmentation of Supply); §4C-3-07 (Water Quality); §4C-3-08 (Underground Water; Limit on Withdrawals); §4C-3-09 (Atmospheric Water).

§2C-2-06 EQUITABLE AND REASONABLE UTILIZATION

Utilization of a transboundary water resource in an equitable and reasonable manner requires taking into account all relevant factors and circumstances, including:

(a) Geographic, hydrographic, hydrological, climatic, ecological and other factors of a natural character;

(b) The social and economic needs of the Parties concerned;

(c) The population dependent on the water resource in each of the Parties;

(d) The effects of the use or uses of the water resources by one Party on other Parties;

(e) Existing and potential uses of the water resource;

(f) Conservation, protection, development and economy of use of the water resource and the costs of measures taken to that effect;

(g) The availability of alternatives, of comparable value, to a particular planned or existing use.

Commentary. This definition appears in the 1997 *Convention on the Law of the Non-Navigational Uses of International Watercourses*, which was adopted under U.N. auspices by all but three nations. This term is wider in scope than the term for "equitable apportionment," providing consideration of many more water uses and programs. For this reason, the term "equitable and reasonable utilization" has been adopted by the committee for the comprehensive water management model. For those Parties within the United States who wish to maintain allegiance to the term "equitable apportionment,: that term should replace the term "equitable and reasonable utilization."

Cross References: §1C-1-02 (purposes and scope of agreement); §3C-2-01 (general powers and duties); §1C-1-04 (coordination and cooperation); §3C-2-10 (meetings, hearings and records); §4C-3-01 (existing rights recognized).

§2C-2-07   FLOOD

"Flood" conditions means conditions resulting from heavy runoff with a frequency specified jointly by the Parties.

Commentary: The flood condition is the opposite of a drought. A large amount of water is to be controlled by facilities of the Commission. The Parties are to agree as to the frequency of the flow of high waters in the Basin. Most of the time, these flows are during periods that exceed the amount of flow that occurs during the years of mean annual precipitation.

Cross-references: §4C-1-03 (comprehensive water management plan); §4C-1-04 (purpose and objectives of comprehensive Water management plan); §4C-1-05 (conditions of comprehensive water management Plan); §4C-1-06 (deviation from comprehensive water management plan); §4C-1-07 (allocation during flood conditions); §4C-1-09 (non-impairment of comprehensive water Management plan by state action); §4C-2-01 (water allocation, generally); §4C-2-03 (basin water sources); §4C-2-06 (watershed management); §4C-3-01 (existing rights recognized); §4C-3-02 (flood protection works)

§2C-2-08   INTERBASIN   TRANSFER

An "interbasin transfer" is any transfer of water, in excess of (____) gallons/liters per day, from one water basin to another.

Commentary: This definition merely makes clear that the term "interbasin transfer" is not limited in any fashion but refers to all transfers from one water basin to another. The provisions regarding interbasin transfers allow regulations to exempt certain small transfers. In many states within the United States, 100,000 gallons per day (378,571 liters per day) are exempt from regulation.

Cross-references: §4C-1-09 (non-impairment of comprehensive water management plan by state action).

§2C-2-09   PARTY OR PARTIES

"Party or Parties" means, unless the text otherwise indicates, those governmental units signatory to this Agreement.

Commentary: Defining the terms in this way avoids the need to include similar language at numerous points throughout the agreement. As a matter of law, it may be unnecessary to state this principle

Cross References: §1C-1-02 (purposes of agreement); §1C-1-03 (0bjectives of agreement); §1C-1-04 (coordination and cooperation); §1C-1-06 (national security); §2C-1-01 (activation of agreement); §2C-1-02 (consent to jurisdiction); §2C-1-03 (duration of agreement); §2C-1-04 (amendments and supplements); §2C-1-05 (powers of sovereign parties; withdrawal); §2C-1-06 (existing agencies); §3C-1-03 (commissioners); §3C-1-04 (status, immunities and privileges); §3C-1-05 (commission organization and staffing); §3C-2-01 (general powers and duties); §3C-2-02 (powers and duties reserved to the commissioners); §3C-2-03 (obligations of the commission); §3C-2-05 (prohibited activities); §3C-2-08 (reports); §3C-2-09 (condemnation proceedings); §3C-2-10 (meetings, hearings and records); §3C-2-11 (tort liability); §3C-3-01 (coordination and cooperation); §3C-3-02 (project costs and evaluation standards); §3C-3-03 (projects of The signatory parties); §3C-3-04 (cooperative services); §4C-1-01 (joint exercise of sovereignty); §4C-1-05 (conditions of comprehensive water management plan); §4C-1-07 (allocation during flood conditions); §4C-2-02 (waters not subject to allocation); §4C-2-05 (allocation to equitable and reasonable uses); §4C-3-01 (existing rights recognized); §4C-3-02 (flood protection works); §4C-3-03 (minimum flows); §4C-3-04 (withdrawals and diversion; protected areas;) §4C-3-05 (water levels protected; §4C-3-07 (water quality); §4C-3-08 (nderground water; limit on withdrawals); §4C-3-09 (atmospheric water); §4C-3-10 (recreation); §6C-1-01 (good Faith implementation); §6C-1-02 (modification of agreement); §6C-1-03 (material breach); §6C-1-04 (alternative dispute resolution); §6C-1-05 (negotiations and consultations); §6C-1-06 (consultations and mediation); §6C-1-07 (arbitration); §6C-1-08 (right to litigate); Annex C-1; Annex C-2.

§2C-2-10 UNDERGROUND WATER

The term "underground water" means water found beneath the ground, regardless of whether flowing through defined channels or percolating through the ground, and regardless of its existence in water table or artesian condition, whether the result of natural or artificial recharge.

Commentary: This definition of "underground water" includes all forms of water in the ground, being equivalent to terms such as "ground water," "groundwater," or similar expressions. It excludes soil (capillary) moisture that might be drawn upon by plants but cannot practically be withdrawn by direct human activity. A somewhat more precise definition is found in the Illinois Water Use Act: water under the ground where the fluid pressure in the pore space is equal to or greater than atmospheric pressure. *See also* Dellapenna, § 6.04; Earl Finbar Murphy, *Quantitative Groundwater,* 3 Waters and Water Rights § 18.02.

Cross-reference: §3C-1-02 (jurisdiction of the commission); §3C-2-03 (obligations of the commission); §4C-2-03 (basin water sources); §4C-3-04, Withdrawals and Diversions; Protected Areas; §4C-3-07, Water Quality; §4C-3-08 (underground water; limit on withdrawals).

§ 2C-2-11 WATERS OF THE BASIN

"Waters of the Basin" shall include all water found within the Basin, whether surface, underground water, or atmospheric water.

Commentary: This definition should be included to make it clear that underground water and atmospheric water are included within the scope of the agreement, if that is the intent of the Parties. With regard to underground water, where current scientific knowledge is available in an aquifer to establish direct interconnectivity between surface and underground water, it is essential that such underground water be included in the Agreement. The technological questions relating to atmospheric water may result in uncertainty regarding its allocation, but to the extent the Parties wish to reach a complete agreement, the matter should be addressed, or recognition given to the fact that the Parties have chosen to reserve that issue for later resolution.

The Parties should also decide whether water imported from other basins should be included within the scope of the agreement. If it is not to be so included, that exclusion should be noted in this section.

Cross References: §1C-1-02 (purposes and scope of agreement); §2C-1-05 (powers of sovereign parties; withdrawal); §2C-2-10 (underground water); §3C-1-02 (jurisdiction of the commission); §3C-2-01 (general powers and duties);§4C-1-01 (purpose and objectives of the comprehensive water management plan); §4C-2-01 (water allocation, generally); §4C-3-01 (existing rights recognized); §4C-3-04 (withdrawals and diversions; protected areas); §4C-3-07 (water quality).

ARTICLE 3C

ADMINISTRATION

Part 1 Administrative Authority

§3C-1-01   COMMISSION   CREATED

(a) The _____ River Commission (hereinafter called the Commission) is hereby created as a body politic and corporate, with succession for the duration of this Agreement, as an agency and instrumentality of the governments of the respective signatory Parties.

(b) The Commission shall develop and effectuate policies for the allocation of the water resources of the _____ River and its tributaries in accordance with this Agreement and may amend the Agreement should changing conditions or extreme events require alteration of the Agreement terms, but only when such changes are agreed to by all Commissioners.

Commentary.   Nearly two-thirds of the interstate water apportionment compacts create a compact commission. *See* Grant, § 46.03. Comprehensive management of a water basin shared by two or more political sovereigns requires joint or communal management. *See* Joseph Dellapenna, *Treaties as Instruments for Managing Internationally Shared-Water Resources: Restricted Sovereignty vs. Community of Property*, 26 Case Western Reserve J. of Int'l Law 27 (1994). To be effective in managing water and precluding conflict, the institutional structure should not only have to embody concepts of cooperative management, but it should also have to be able to:, (1) determine the facts of water use in each nation; (2) resolve disputes across political boundaries; (3) guide responses to unusual temporary water shortages; (4) regulate or design long-term solutions; and (5) enforce its decision. *See* e.g. U.N. Department of Economic and Social Affairs, Natural Resources Water Series No. 1, *Management of Int'l Water Resources, Institutional and Legal Aspects*, Report of the Panel of Experts on the Legal and Institutional Aspects of International Water Resources Development, U.N. Doc. St/ESA/5; David Le Marquand, *Politics of International Basin Cooperation and*

Management, in Water in a Developing World (Albert E. Utton & Ludwick A. Teclaff, eds., 1978). The organizational structure of the Commission should be constituted according to the specific geographical and hydrological characteristics of the shared water resource and the political structures of the parties involved. "What works for wealthy nations may not work for developing countries," and, in some shared water situations, cultural differences (e.g., the Jordan River) may require different management structures than used in those situations involving similar cultures (e.g., the Rhine River). *See* Arthur E. Williams, *The Role of Technology in Sustainable Development* in Water Resources Administration in the United States (M. Reuss, ed., 1993}.

Sterner states that "there must be an appropriate legal structure and set of institutions that define property rights and establish the framework within which an environmental authority can function." (Thomas Sterner, Economic Policies for Sustainable Development, (1994).

Cross-references: §2C-2-02 (basin); §3C-1-02 (jurisdiction of the commission); §3C-1-03 (commissioners); §3C-1-04 (status, immunities and privileges); §3C-1-05 (commission organization and staffing); §3C-1-06 (rules of procedures); §3C-1-07 (commission administration); §3C-2-01 (general powers and duties); §3C-2-02 (powers and duties reserved to the commissioners); §3C-2-03 (obligations of the commission); §3C-2-04 (regulations; enforcement); §3C-2-05 (prohibited activities); §3C-2-06 (referral and review); §3C-2-07 (advisory committees); §3C-2-08 (reports); §3C-2-09 (condemnation proceedings); §3C-2-10 (meetings, hearings and records); §3C-2-11 (tort liability); §3C-3-01 (coordination and cooperation); §3C-3-02 (project costs and evaluation standards); §3C-3-03 (projects of the signatory parties); §3C-3-04 (cooperative services).

Similar Agreements: *Klamath River Basin Compact,* 71 Stat. 497 (1957); *Delaware River Basin Compact,* Pub. L. 87-328, 75 Stat. 688 (1961); *Susquehanna River Basin Compact,* Pub. L. No. 91-575, 84 Stat. 1509 (1970).; *Apalachicola-Chattahoochee-Flint River Basin Compact,* O.C.G.A. 12-10-100 (1997); *Alabama-Coosa-Tallapoosa River Basin Compact,*O.C.G.A. 12-10-110 (1997); *Treaty between the United States and Great Britain relating to Boundary Waters, and Questions arising between the United States and Canada,* 36 Stat. 2451 (1909); *Treaty between the United States of America and Mexico, Utilization of Waters of the Colorado and Tijuana rivers and of the Rio Grande,* 59 Stat. 1219 (1945); Indus Water Treaty, 419 UNTS 126 (1960); *The North American Agreement on Environmental Cooperation between the Government of the United States of America, the Government of Canada, and the Government of the United Mexican States,* 32 I.L.M. 1480 (1993); *Agreement on Cooperation for the Sustainable Development of the Mekong River Basin,* 34 ILM 864 (1995).

§3C-1-02   JURISDICTION OF THE COMMISSION

(a) The Commission shall have, exercise and discharge its functions, powers and duties within the limits of the Basin, except that it may in its discretion act outside the Basin whenever such action may be necessary or convenient to effectuate its powers or duties within the Basin. The Commission shall exercise such power outside the Basin only upon the consent of the Party in which it proposes to act.

(b) Those sources of underground water that have a direct connection and recharge capability to surface waters of the Basin shall be included in the water allocations of this Agreement. In the event such underground water aquifers are or have the potential of being used by parties not signatory to this Agreement, the Commission will enter into consultations and negotiations with such parties to reach an agreement as to the allocation of the interconnected underground water.

Commentary. This provision describes the geographic and hydrologic jurisdiction of the Commission. Similar provisions appear in most agreements in order to clarify and define the limits of supranational authority, although conjunctive management has not been the norm. *See* Hayton, R.D. and A.E. Utton, Transboundary Groundwaters: The Bellagio Draft Treaty (1989).

Cross-references: §2C-2-02 (basin); §2C-2-11 (waters of the basin); §3C-1-01 (commission created); §3C-1-03 (commissioners); §3C-1-04 (status, immunities and privileges); §3C-1-05 (commission organization and staffing); §3C-1-06 (rules of procedures); §3C-1-07 (commission administration); §3C-2-01 (general powers and duties); §3C-2-02 (powers and duties reserved to the commissioners); §3C-2-03 (obligations of the commission); §3C-2-04 (regulations; enforcement); §3C-2-05 (prohibited activities); §3C-2-06 (referral and review); §3C-2-07 (advisory committees); §3C-2-08 (reports); §3C-2-09 (condemnation proceedings); §3C-2-10 (meetings, hearings and records); §3C-2-11 (tort liability); §3C-3-01 (coordination and cooperation); §3C-3-02 (project costs and evaluation standards); §3C-3-03 (projects of the signatory parties); §3C-3-04 (cooperative services).

Similar Agreements: *Delaware River Basin Compact*, Pub. L. 87-328, 75 Stat. 688 (1961); *Susquehanna River Basin Compact*, Pub. L. No. 91-575, 84 Stat. 1509 (1970); *Treaty between the United States and Great Britain relating to Boundary Waters, and Questions arising between the United States and Canada*, 36 Stat. 2451 (1909); *Treaty Respecting Utilization of Water in the Colorado and Tijuana Rivers and of the Rio Grande*, 59 Stat. 1219 (1945); *Agreement on the Cooperation for the Sustainable Development of the Mekong River Basin*, 34 ILM 864 (1995).

§3C-1-03   COMMISSIONERS

(a) The Commission shall be governed by a Board of Commissioners, consisting of the Governor or other Chief Executive Officer of the signatory Parties.

(b) Each Commissioner may appoint an alternate to act in his place and stead, with authority to attend all meetings of the Commission, and with power to vote in the absence of the member. Unless otherwise provided by law of the signatory Party for which he is appointed, each alternate shall serve during the term of the member appointing him, subject to removal at the pleasure of the member. In the event of a vacancy in the office of alternate, it shall be filled in the same manner as an original appointment for the unexpired term only.

(c) All matters requiring a policy decision affecting the substance of the Agreement shall be decided upon by the Commissioners and unanimity shall be required. For all other matters, the rule of decision shall be simple majority. Each member shall be entitled to one vote on all matters that may come before the Commission. No action by the Commissioners shall be taken at any meeting unless a majority of the membership shall vote in favor thereof. The federal representative shall have no vote.

Commentary. The membership of the typical commission in the U.S. includes one or more members from each Party plus a federal representative. The federal representative usually has no vote. *See* Grant, §46.03. The DRBC and SRBC are notable exceptions. This provision recognizes that the responsibility (and authority) for all policy decisions or decisions affecting the substance of the Agreement remains with the principal executive*s* office of the respective Parties.

Cross-references: §2C-2-09 (party or parties); §3C-1-01 (commission created); §3C-1-02 (jurisdiction of the commission); §3C-1-04 (status, immunities and privileges); §3C-1-05 (commission organization and staffing); §3C-1-06 (rules of procedures); §3C-1-07 (commission administration); §3C-2-01 (general powers and duties); §3C-2-02 (powers and duties reserved to the commissioners); §3C-2-03 (obligations of the commission); §3C-2-04 (regulations; enforcement); §3C-2-05 (prohibited activities); §3C-2-06 (referral and review); §3C-2-07 (advisory committees); §3C-2-08 (reports); §3C-2-09 (condemnation proceedings); §3C-2-10 (meetings, hearings and records); §3C-2-11 (tort liability); §3C-3-01 (coordination and cooperation); §3C-3-02 (project costs and evaluation standards); §3C-3-03 (projects of the signatory parties); §3C-3-04 (cooperative services).

Similar Agreements: *Klamath River Basin Compact*, 71 Stat. 497 (1957); *Delaware River Basin Compact*, Pub. L. 87-328, 75 Stat. 688 (1961); *Susquehanna River Basin Compact*, Pub. L. No. 91-575, 84 Stat. 1509 (1970); *Apalachicola-Chattahoochee-Flint River Basin Compact*, O.C.G.A. 12-10-100 (1997); *Alabama-Coosa-Tallapoosa River Basin Compact*, O.C.G.A. 12-10-110 (1997); *Convention on the Protection and Use of Transboundary Watercourses and International Lakes*, 31 I.L.M. 1312 (1992).

§3C-1-04 STATUS, IMMUNITIES AND PRIVILEGES (Optional)

To enable the Commission to fulfill its purpose and the functions with which it is entrusted, the status, immunities and privileges set forth in this Article shall be accorded to the Commission in the territories of each Party.

(a) The Commission, its property and its assets, wherever located, and by whomsoever held, shall enjoy the same immunity from suit and every form of judicial process as is enjoyed by the Parties, except to the extent that the Commission may expressly waive its immunity for the purposes of any proceedings or by the terms of any contract.

(b) Property and assets of the Commission, wheresoever located and by whomsoever held, shall be considered public property and shall be immune from search, requisition, confiscation, expropriation or any other form of taking or foreclosure by executive or legislative action.

(c) To the extent necessary to carry out the purpose and functions of the Commission and to conduct its operations in accordance with this Agreement, all property and other assets of the Commission shall be free from restrictions, regulations, controls and moratoria of any nature affecting the implementation of this agreement, except as may otherwise be provided in this Agreement.

(d) The official communications of the Commission shall be accorded by each Party the same treatment that it accords to the official communications of the other Parties.

(e) (Optional, for international use only) The Commissioners and other personnel engaged directly in the affairs of the Commission shall have the following privileges and immunities:

> (1) Immunity from legal process with respect to acts performed by them in their official capacity except when the Commission expressly waives this immunity.
>
> (2) When not citizens of one of the signatory Parties, the same immunities from immigration restrictions, alien registration requirements and national service obligations and the same facilities as regards exchange provisions as are accorded by each Party to the representatives, officials, and employees of comparable rank of the other Party; and
>
> (3) The same privileges in respect of traveling and facilities as are accorded by each Party to representatives, officials, and employees of comparable rank of the other Party.

(f) The Commission, its property, other assets, income, and the operations it carries out pursuant to this Chapter shall be immune from all state taxation. The Commission shall also be immune from any obligation relating to the payment, withholding or collection of any tax or customs duty. No state tax shall be levied on or in respect of salaries and benefits paid by the Commission to officers or staff of the Commission who are not local citizens.

(g) Each Party, in accordance with its juridical system, shall take such action as is necessary to make effective in its own territories the principles set forth in this Article, and shall inform the Commission of the action which it has taken on the matter.

Commentary: This provision provides the Commissioners and their personnel with the same legal protections that normally exist for governmental officials of the Parties.

Cross-references: §1C-1-01 (general policies); §1C-1-02 (purposes of agreement); §1C-1-03 (objectives of agreement); §2C-2-09 (party or parties); §3C-1-01 (commission created); §3C-1-02 (jurisdiction of the commission); §3C-1-03 (commissioners); §3C-1-05 (commission organization and staffing); §3C-1-06 (rules of procedures); §3C-1-07 (commission administration); §3C-2-01 (general powers and duties); §3C-2-02 (powers and duties reserved to the commissioners); §3C-2-03 (obligations of the commission).

Similar Agreements: *Agreement between the Governments of the United States of America and the Government of the United Mexican States Concerning the Establishment of a Border Environment Cooperation Commission and the North American Development Bank*, 19 U.S.C. §§ 3473 (1993).

§3C-1-05 COMMISSION ORGANIZATION AND STAFFING

(a) The Commission shall meet and organize at _____ promptly after the members thereof are appointed, and when organized the Commission may fix such times and places for its meetings as may be necessary, subject at all times to special call or direction by the two Parties. Each Commissioner, upon the first joint meeting of the Commission after his appointment, shall, before proceeding with the work of the Commission, make and subscribe a solemn declaration in writing that he will faithfully and impartially perform the duties imposed upon him under this Agreement, and such declaration shall be entered on the records of the proceedings of the Commission.

(b) The respective Commissioners may each appoint a secretary, and these shall act as joint secretaries of the Commission at its joint sessions, and the Commission may employ professional and administrative personnel from time to time as it may deem advisable. The salaries and personal expenses of the Commission and of the secretaries shall be paid by their respective Governments, and all reasonable and necessary joint expenses of the Commission, incurred by it, shall be paid in equal portions by the Parties unless otherwise stipulated in this agreement.

Commentary: This provision provides the authority and instructions for the organization and initiation of Commission undertakings. The Parties may prefer to apportion joint expenses according to benefits received or some other economic algorithm.

Cross-references: §1C-1-01 (general policies); §1C-1-02 (purposes of agreement); §1C-1-03 (objectives of agreement); §2C-2-09 (party or parties); §3C-1-01 (commission created).

Similar Agreements: *Treaty between the United States and Great Britain relating to Boundary Waters*, 36 Stat. 2451 (1909); *Convention Between the Federal Republic of Germany and the*

Czech and Slovak Federal Republic and the European Economic Community on the International Commission for the Protection of the Elbe, International Environmental Law, Multilateral Agreements, 976:90/1 (1990).

§3C-1-06 RULES OF PROCEDURES

(a) **The Commission shall adopt its own Rules of Procedures.**

(b) **The Commission may seek technical advisory services as it deems necessary.**

Commentary: Except when necessary for policy reasons, the Agreement should not bind the Commission to specific procedural requirements.

Cross-references: §3C-1-01 (commission created); §3C-1-02 (jurisdiction of the commission); §3C-1-03 (commissioners); §3C-1-04 (status, immunities and privileges); §3C-1-05 (commission organization and staffing); §3C-1-07 (commission administration); §3C-2-01 (general powers and duties); §3C-2-02 (powers and duties reserved to the commissioners); §3C-2-03 (obligations of the commission); §3C-2-04 (regulations; enforcement); §3C-2-05 (prohibited activities); §3C-2-06 (referral and review); §3C-2-07 (advisory committees); §3C-2-08 (reports); §3C-2-09 (condemnation proceedings); §3C-2-10 (meetings, hearings and records); §3C-2-11 (tort liability); §3C-3-01 (coordination and cooperation); §3C-3-02 (project costs and evaluation standards); §3C-3-03 (projects of the signatory parties); §3C-3-04 (cooperative services).

Similar Agreements: *Klamath River Basin Compact*, 71 Stat. 497 (1957); Delaware *River Basin Compact*, Pub. L. 87-328, 75 Stat. 688 (1961); *Susquehanna River Basin Compact*, Pub. L. No. 91-575, 84 Stat. 1509 (1970); *The North American Agreement on Environmental Cooperation between the Government of the United States of America, the Government of Canada, and the Government of the United Mexican States*, 32 I.L.M. 1480 (1993); *Agreement on the Cooperation for the Sustainable Development of the Mekong River Basin*, 34 ILM 864 (1995); *Convention on the Law of the Non-Navigational Uses of International Watercourses*, United Nations Document A/51/869 (1998).

§3C-1-07 COMMISSION ADMINISTRATION

(a) **Supervision and management of the routine and customary affairs of the Commission shall be vested in a Commission Administration consisting of an Executive Director and such additional officers, deputies and assistants as the Commissioners may determine. The Executive Director shall be appointed and may be removed by the affirmative vote of a majority of the full membership of the Commissioners. All other officers and employees shall be appointed by the Executive Director and confirmed by the Commissioners under such rules of procedure as the Commission may determine.**

(b) Except for the express limitations provided in this Agreement, implementation and administration of the Agreement terms shall be vested in the Commission Officers appointed by the Commissioners. The Commission Officers shall report directly to the Commission.

(c) In the appointment and promotion of Commission Officers and employees, no political, racial, gender, religious or residence test or qualification shall be permitted or given consideration, but all such appointments and promotions shall be solely on the basis of merit and fitness. Any officer or employee of the Commission who is found by the Commissioners to be guilty of a violation of this section shall be removed from office.

Commentary. The Commission Administration provides the day-to-day administrative management of the comprehensive plan for the Commission. While all substantive policy-making decisions involving the Agreement rightly remain with the Commissioners, the Commission's purpose is to carry out the technical facets involved with Agreement implementation.
  The Parties may wish to establish an Commission membership that equitably allocates officers and employees among citizens of the various governments. In such a case, the references to political and residency tests can be eliminated. Non-discrimination on the basis of race, gender or religion is strongly encouraged however.

Cross-references: §3C-1-01 (commission created); §3C-1-02 (jurisdiction of the commission); §3C-1-03 (commissioners); §3C-1-04 (status, immunities and privileges); §3C-1-05 (commission organization and staffing); §3C-1-06 (rules of procedures); §3C-2-01 (general powers and duties); §3C-2-02 (powers and duties reserved to the commissioners); §3C-2-03 (obligations of the commission); §3C-2-04 (regulations; enforcement); §3C-2-05 (prohibited activities); §3C-2-06 (referral and review); §3C-2-07 (advisory committees); §3C-2-08 (reports); §3C-2-09 (condemnation proceedings); §3C-2-10 (meetings, hearings and records); §3C-2-11 (tort liability); §3C-3-01 (coordination and cooperation); §3C-3-02 (project costs and evaluation standards); §3C-3-03 (projects of the signatory parties).

Similar Agreements: *Klamath River Basin Compact*, 71 Stat. 497 (1957); *Delaware River Basin Compact*, Pub. L. 87-328, 75 Stat. 688 (1961); *Susquehanna River Basin Compact*, Pub. L. No. 91-575, 84 Stat. 1509 (1970); *Apalachicola-Chattahoochee-Flint River Basin Compact*, O.C.G.A. 12-10-100 (1997); *Alabama-Coosa-Tallapoosa River Basin Compact*, O.C.G.A. 12-10-110 (1997); *Agreement on the Cooperation for the Sustainable Development of the Mekong River Basin*, 34 ILM 864 (1995).

Part 2 Powers and Duties

§3C-2-01 GENERAL POWERS AND DUTIES

The Commission as a corporate body shall:

(a) Allocate the waters of the Basin among the Parties, in accordance with the doctrine of equitable and reasonable utilization, to and among the states signatory to this Agreement and to and among their respective political subdivisions, and to impose conditions, obligations and release requirements related thereto;

(b) Develop, implement and effectuate plans and projects for the use of the water of the Basin for the purposes of economical growth and development, maintenance of adequate public health and safety, and environmental protection;

(c) Provide administration and coordination of the Agreement implementation;

(d) Conduct activities consistent with its functions so as to further the conservation and enhancement of natural beauty and the sustainability of the environment;

(e) Coordinate the collection, compilation and analysis of data and information of hydrologic, environmental and economic importance within the Basin;

(f) Respond on an emergency basis to changing conditions when time constraints do not allow review and analysis by the Commissioners;

(g) Report (quarterly/semi-annually/annually) to the Parties on the implementation of the Agreement;

(h) Conduct active public involvement programs to ensure the needs of all stakeholders are considered.

(i) Prepare annual budgets for implementation of the Agreement;

(j) Identify and recommend to the Commissioners appointments to the Commission administration and any review panel;

(k) Establish standards for planning, design and operation of all water resources projects and facilities in the Basin;

(l) Conduct and sponsor research on water resources issues that may arise during the life of the Agreement;

(m) Prepare, publish and disseminate information and reports with respect to the water problems of the Basin and for the presentation of the needs, resources and policies of the Basin to the Commissioners and the executive and legislative branches of the signatory Parties;

(n) Plan, manage, budget and allocate financial resources necessary to effectuate the purposes of this Agreement; and

(o) Exercise such othe powers as may be delegated to it by this Agreement or otherwise pursuant to law, and have and exercise all powers necessary or convenient to carry out its express powers or powers which may be reasonably implied therefrom.

Commentary. This list of powers and duties should be sufficient to adequately manage the implementation of the Agreement. Care should be taken when adding powers to ensure that they do not include substantive policy-making powers that may conflict with state sovereignty issues. Additionally, the stated powers should not be so detailed as to restrict the management of the inherently variable conditions of water resources.

Cross-references: §1C-1-01 (general policies); §1C-1-02 (purposes of agreement); §1C-1-03 (objectives of agreement); §2C-2-02 (basin); §2C-2-06 (equitable and reasonable apportionment); §2C-2-09 (party or parties); §2C-2-11 (waters of the basin); §3C-2-02 (powers and duties reserved to the commissioners); §3C-2-03 (obligations of the commission); §3C-2-04 (regulations; enforcement); §3C-2-05 (prohibited activities).

Similar Agreements: *Klamath River Basin Compact*, 71 Stat. 497 (1957); *Delaware River Basin Compact*, Pub. L. 87-328, 75 Stat. 688 (1961); *Susquehanna River Basin Compact*, Pub. L. No. 91-575, 84 Stat. 1509 (1970); *Delaware River Basin Compact*, Pub. L. 87-328, 75 Stat. 688 (1961); *Susquehanna River Basin Compact*, Pub. L. No. 91-575, 84 Stat. 1509 (1970); Apalachicola-*Chattahoochee-Flint River Basin Compact*, O.C.G.A. 12-10-100 (1997); *Alabama-Coosa-Tallapoosa River Basin Compact*, O.C.G.A. 12-10-110 (1997); *Treaty between the United States of America and Mexico, Utilization of Waters of the Colorado and Tijuana rivers and of the Rio Grande*, 59 Stat. 1219 (1945); *The North American Agreement on Environmental Cooperation between the Government of the United States of America, the Government of Canada, and the Government of the United Mexican States*, 32 I.L.M. 1480 (1993); *The Declaration of Principles on Interim Self-Government Arrangements, Sep. 13, 1993, Israel.-PLO*, 32 I.L.M. 1525 (1993); *Agreement on the Cooperation for the Sustainable Development of the Mekong River Basin*, 34 ILM 864 (1995).

§3C-2-02 POWERS AND DUTIES RESERVED TO THE COMMISSIONERS

The Commissioners shall supervise and control the implementation of the provisions of the Agreement. The Commissioners may delegate certain powers but shall retain the following:

(a) To appoint the Executive Director of the Commission and confirm the appointment of other Commission Officers designated to administer the Agreement provisions;

(b) To adopt annual budgets prepared for Agreement administration;

(c) To approve all plans and capital projects developed by the signatory Parties relating to the water resources of the Basin;

(d) (Optional) To approve all fees and assessments levied by the Commission;

(e)) To adopt and promote uniform and coordinated policies for water conservation, control, use and management in the Basin and to approve the planning, development and financing of water resources projects according to such plans and policies;

(f) To require the maintenance of faithful records of all meetings and decision-making activities of the Commission and provide public access to those records (within [___] calendar days) of the meeting or activities taking place;

(g) To conduct public meetings (once/twice) yearly concerning the activities of the Commission and provide opportunity for all interest groups and the generally public to express their views.

(h) To determine the character of and the necessity for its obligations and expenditures and the manner in which they shall be incurred, allowed, and paid subject to any provisions of law specifically applicable to agencies or instrumentalities created by Agreement;

(i) To provide for the internal organization and administration of the Commission;

(j) To appoint the principal executive officers of the Commission and delegate to and allocate among them administrative functions, powers and duties;

(k) To create and abolish offices, employment opportunities and positions as it deems necessary for the purposes of the Commission, and subject to the provisions of this article, fix and provide for the qualification, appointment, removal, term, tenure, compensation, pension and retirement rights of its officers and employees; and

(l) (Optional) To let and execute contracts over ($500,000) to carry out the purposes of this Agreement

Commentary. These powers and duties of the Commissioners are the minimum necessary to effectively administer a Comprehensive Water Management Plan while ensuring that all stakeholders have a channel to express their needs. Although the list can be expanded, care must be taken to further limit any infringement of national sovereignty that may cause the Agreement to be breached.

Subparagraph (d) containes an optional provision for approval of a power granted below to the Commission itself to levy fees and assessments to cover costs of specific operations and maintenance. Its inclusion is appropriate only when the power to levy fees and assessments (see *Delaware River Basin Compact*, Pub. L. 87-328, 75 Stat. 688 (1961)) has been given to the Commission (see §3C-2-03(e) below).

Tlist of powers also includes the requirement for approval by the Commissioners of contracts in excess of a certain amount, to be determined by the Parties.

Cross-references: §1C-1-01 (general policies); §1C-1-02 (purposes of agreement); §1C-1-03 (objectives of agreement); §2C-2-02 (basin); §2C-2-09 (party or parties); §2C-2-10 (underground water); §3C-2-03 (obligations of the commission); §3C-2-04 (regulations; enforcement); §3C-2-05 (prohibited activities).

Similar Agreements: *Delaware River Basin Compact*, Pub. L. 87-328, 75 Stat. 688 (1961); *Susquehanna River Basin Compact*, Pub. L. No. 91-575, 84 Stat. 1509 (1970); *Apalachicola-Chattahoochee-Flint River Basin Compact*, O.C.G.A. 12-10-100 (1997); *Alabama-Coosa-Tallapoosa River Basin Compact*, O.C.G.A. 12-10-110 (1997); *Treaty between the United States of America and Mexico, Utilization of Waters of the Colorado and Tijuana rivers and of the Rio Grande*, 59 Stat. 1219 (1945); Treaty between the United States of America and Mexico, *The North American Agreement on Environmental Cooperation between the Government of the United States of America, the Government of Canada, and the Government of the United Mexican States*, 32 I.L.M. 1480 (1993).

§3C-2-03   POWERS AND DUTIES OF THE COMMISSION (Optional)

The Commission, for the purposes of this Agreement, may:

(a) Enter into contracts, sue and be sued in all courts of competent jurisdiction;

(b) Receive and accept such payments, appropriations, grants, gifts, loans, advances and other funds, properties and services as may be transferred or made available to it by any signatory Party or by any other public or private corporation or individual, and enter into agreements to make reimbursement for all or part thereof;

(c) Provide for, acquire and adopt detailed engineering, administrative, financial and operating plans and specifications to effectuate, maintain or develop any facility or project;

(d) Control and regulate the use of facilities owned or operated by the Commission;

(e) (Optional) Assess on an annual basis or otherwise the cost of operations and maintenance upon water users or any classification of them specially benefited thereby to a measurable extent, provided that no such assessment shall exceed the actual benefit to any water user and such assessment shall follow the procedure prescribed by law for the assessment of fees for governmental services and shall be subject to judicial review in any court of competent jurisdiction,

(f) Acquire, own, operate, maintain, control, sell and convey real and personal property and any interest therein by contract, purchase, lease, license, mortgage or otherwise as it may deem necessary for any project or facility, including any and all appurtenances thereto necessary, useful or convenient for such ownership, operation, control, maintenance or conveyance;

(g) Provide for, construct, acquire, operate and maintain dams, reservoirs and other facilities for utilization of surface and water resources, and all related structures, appurtenances and equipment in the Basin and its tributaries and at such off-river sites as it may find appropriate, and may regulate and control the use thereof.

(h) Have and exercise all corporate powers essential to the declared objects and purposes of the Commission.

Commentary. This provision allows the Commission to operate without the day-to-day decision-making by the Commisioners themselves. However, although the powers and duties listed in this section assist in effective implementation of the Agreement, the provision is optional and may be deleted or expanded, according to the specific geopolitical situation.

Subparagraph (e) containes an optional provision for the levy of fees and assessments to cover costs of specific operations and maintenance. (See *Delaware River Basin Compact*, Pub. L. 87-328, 75 Stat. 688 (1961)) Its inclusion is appropriate only if this power is subkect to the approval of the Commissioners (see §3C-2-02(d) above).

Cross-references: §1C-1-01 (general policies); §1C-1-02 (purposes of agreement); §1C-1-03 (objectives of agreement); §2C-2-02 (basin); §2C-2-09 (party or parties); §2C-2-10 (underground water); §3C-2-01 (general powers and duties); §3C-2-02 (powers and duties reserved to the commissioners); §3C-2-04 (regulations; enforcement); §3C-2-05 (prohibited activities).

Similar Agreements: *Delaware River Basin Compact*, Pub. L. 87-328, 75 Stat. 688 (1961); *Susquehanna River Basin Compact*, Pub. L. No. 91-575, 84 Stat. 1509 (1970).

§3C-2-04 REGULATIONS; ENFORCEMENT

The Commission may:

(a) Make and enforce reasonable rules and regulations for the initiation, application and enforcement of this Agreement; and it may adopt and enforce practices and schedules for or in connection with the use, maintenance and administration of projects and facilities it may own or operate and any product or service rendered thereby; provided that any rule or regulation, other than one which deals solely with the internal management of the Commission, shall be adopted only after public hearing and shall not be effective unless and until filed in accordance with the law of the respective signatory Parties applicable to administrative rules and regulations;

(b) Provide regulations for water management in the Basin, and

(c) Designate any officer, agent or employee of the Commission to be an investigator or watchman, and such person shall be vested with the police powers of of the jurisdiction in which he is duly assigned to perform his duties.

Commentary. Any successful comprehensive plan requires detailed regulations. This section provides the Commission with the power to establish and enforce regulations necessary for Agreement implementation. The provision was modeled on a similar provision in the DRBC and SRBC with several changes.

Cross-references: §1C-1-01 (general policies); §1C-1-02 (purposes of agreement); §1C-1-03 (objectives of agreement); §2C-2-06 (equitable and reasonable apportionment); §3C-2-01 (general powers and duties); §3C-2-02 (powers and duties reserved to the commissioners); §3C-2-03 (obligations of the commission); §3C-2-05 (prohibited activities).

Similar Agreements: *Delaware River Basin Compact*, Pub. L. 87-328, 75 Stat. 688 (1961); *Susquehanna River Basin Compact*, Pub. L. No. 91-575, 84 Stat. 1509 (1970).

§3C-2-05   PROHIBITED   ACTIVITIES (Optional)

(a) No Commissioner or Commission Officer or employee shall:

(1) Be financially interested, either directly or indirectly, in any contract, sale, purchase, lease or transfer of real or personal property in which the Commission is involved;

(2) Solicit or accept money or any other thing of value, either directly or indirectly, in addition to the compensation or expenses paid him by the Commission for services performed within the scope of his official duties;

(3) Offer money or any thing of value for or in consideration of obtaining an appointment, promotion or privilege in his employment with the Commission.

(b) Any officer or employee who shall willfully violate any of the provisions of this section shall forfeit his office or employment.

(c) Any contract or agreement knowingly made in contravention of this section is void.

(d) Officers and employees of the Commission shall be subject to such criminal and civil sanctions for misconduct in office as may be imposed by [federal law and] the law of the signatory Party in which such misconduct occurs.

Commentary. Although optional, inclusion of this section provides professionalism of the administration of the Agreement and guards against corruption in its enforcement.

Cross-references: §1C-1-01 (general policies); §1C-1-02 (purposes of agreement); §1C-1-03 (objectives of agreement); §2C-2-09 (party or parties); §3C-2-01 (general powers and duties); §3C-2-02 (powers and duties reserved to the commissioners); §3C-2-03 (obligations of the commission); §3C-2-04 (regulations; enforcement).

Similar Agreements: *Delaware River Basin Compact*, Pub. L. 87-328, 75 Stat. 688 (1961); *Susquehanna River Basin Compact*, Pub. L. No. 91-575, 84 Stat. 1509 (1970).

§3C-2-06 COMMISSION APPROVAL OF WATER RESOURCE PROJECTS (Optional)

No project having a substantial effect on the water resources of the Basin shall hereafter be undertaken by any person, corporation or governmental authority unless it shall have been first submitted to and approved by the Commission. The Commission shall approve a project whenever it finds and determines that such project would not substantially impair or conflct with the comprehensive plan and may modify and approve as modified, or may disapprove any such project whenever it finds and determines that the project would substantially impair or conflict with such plan. The Commission shall provide by regulation for the procedure of submission, review and consideration of projects, and for its determinations pursuant to this section.

Commentary. This section establishes the primacy of the Commission in water resource planning, providing it with the sole authority to approve projects that substantially affect the Basin's water resources. Without this restriction, the purposes of the plan may be thwarted. The determination of what constitutes "substantial impairment or conflict" is case specific. Although optional, its inclusion is strongly advised.

Cross-references: §2C-1-06 (existing agencies); §2C-2-07 (flood); §3C-3-02 (project costs and evaluation standards); §3C-3-03 (projects of the signatory parties); §3C-3-04 (cooperative services); §4C-1-01 (joint exercise of sovereignty); §4C-1-01 (joint exercise of sovereignty); §4C-1-02 (interrelationship of water resources); §4C-1-03 (comprehensive water management plan); §4C-1-04 (purpose and objectives of comprehensive water management plan); §4C-1-05 (conditions of comprehensive water management Plan); §4C-1-06 (Deviation from Comprehensive Water Management plan); §4C-1-07 (allocation during flood conditions); §4C-1-

08 (Allocation during drought conditions); §4C-1-09 (non-impairment of comprehensive water management plan by state action); §4C-2-01 (water allocation, generally); §4C-2-02 (waters not subject to allocation); §4C-2-03 (basin water sources); §4C-2-04 (basin water demands and needs); §4C-2-05 (allocation to equitable and reasonable uses); §4C-2-06 (watershed management); §4C-3-01 (existing rights recognized); §4C-3-02 (flood protection works); §4C-3-03 (minimum flows); §4C-3-04 (withdrawals and diversion; protected areass); §4C-3-05 (water levels protected); §4C-3-06 (augmentation of supply); §4C-3-07 (water quality); §4C-3-08 (underground water; limit on withdrawals); §4C-3-09 (atmospheric water); §4C-3-10 (recreation).

Similar Agreements: *Delaware River Basin Compact*, Pub. L. 87-328, 75 Stat. 688 (1961); *Susquehanna River Basin Compact*, Pub. L. No. 91-575, 84 Stat. 1509 (1970); *Treaty between the United States of America and Mexico, Utilization of Waters of the Colorado and Tijuana rivers and of the Rio Grande*, 59 Stat. 1219 (1945).

§3C-2-07 ADVISORY COMMITTEES (Optional)

The Commission may constitute and empower advisory committees, which may be comprised of representatives of the public and of agencies and officials of the Parties or any sub-division thereof, water-using industries, water-interest groups, and academic experts in related fields.

Commentary. Although optional, inclusion of this section allows the Commission to seek independent advice and counsel regarding water issues within the Basin, thereby assuring maximum technical competence being applied to the comprehensive water management.

Cross-references: §3C-1-05 (commission organization and staffing); §3C-1-06 (rules of procedures); §3C-1-07 (commission administration); §3C-2-01 (general powers and duties); §3C-2-02 (powers and duties reserved to the commissioners); §3C-2-03 (obligations of the commission).

Similar Agreements: *Delaware River Basin Compact*, Pub. L. 87-328, 75 Stat. 688 (1961); *Susquehanna River Basin Compact*, Pub. L. No. 91-575, 84 Stat. 1509 (1970).

§3C-2-08 REPORTS (Optional)

The Commission shall make and publish an Annual Report to the Commissioners and the legislative bodies of the signatory Parties and to the public reporting on its programs, operations and finances. It may also prepare, publish and distribute such other public reports and informational materials as it may deem necessary or desirable.

Commentary. Although optional, inclusion of this section is strongly advised. It reinforces the public nature of the Agreement administration. Perhaps more importantly, such reporting provides for public accountability of the Commission and assures maximum effective management of the water sources and uses.

Cross-references: §1C-1-01 (general policies); §1C-1-02 (purposes of agreement); §1C-1-03 (objectives of agreement); §2C-2-09 (party or parties); §3C-1-01 (commission created); §3C-1-02 (jurisdiction of the commission); §3C-3-02 (project costs and evaluation standards); §3C-3-03 (projects of the signatory parties); §4C-1-03 (comprehensive water management plan); §4C-1-04 (purpose and objectives of comprehensive water management plan); §4C-1-05 (conditions of comprehensive water management Plan); §4C-1-06 (Deviation from Comprehensive Water Management plan); §4C-1-07 (allocation during flood conditions); §4C-1-08 (Allocation during drought conditions); §4C-1-09 (non-impairment of comprehensive water management plan by state action); §4C-2-01 (water allocation, generally); §4C-2-02 (waters not subject to allocation); §4C-2-03 (basin water sources); §4C-2-04 (basin water demands and needs); §4C-2-05 (allocation to equitable and reasonable uses); §4C-2-06 (watershed management); §4C-3-01 (existing rights recognized); §4C-3-02 (flood protection works); §4C-3-03 (minimum flows); §4C-3-04 (withdrawals and diversion; protected areas); §4C-3-05 (water levels protected); §4C-3-06 (augmentation of supply); §4C-3-07 (water quality); §4C-3-08 (underground water; limit on withdrawals); §4C-3-09 (atmospheric water); §4C-3-10 (recreation).

Similar Agreements: *Delaware River Basin Compact*, Pub. L. 87-328, 75 Stat. 688 (1961); *Susquehanna River Basin Compact*, Pub. L. No. 91-575, 84 Stat. 1509 (1970).

§3C-2-09 CONDEMNATION PROCEEDINGS (Optional)

(a) **The Commission shall have the power to acquire by condemnation the fee or any lesser interest in lands, lands lying under water, development rights in land, riparian rights, water rights, waters and other real or personal property within the Basin for any project or facility authorized pursuant to this Agreement. This grant of power of eminent domain includes but is not limited to the power to condemn for the purposes of this Agreement any property already devoted to a public use, by whomsoever owned or held, other than property of a signatory Party. Any condemnation proceeding of any property or franchises owned or used by a municipal or privately owned public utility, unless the affected public utility facility is to be relocated or replaced, shall be subject to the authority of such state board, Commission or other body as may have regulatory jurisdiction over such public utility.**

(b) **Such power of condemnation shall be exercised in accordance with the provisions of any law applicable to the jurisdiction in which the property is located.**

(c) **Nothing in this Agreement authorizes the taking of any existing vested property right in the use of water except for just compensation, in accordance with the internal laws of the Party in which the property or usafructary right exists.**

Commentary. This provision is primarily concerned with the condemnation of real property that may be required for projects and facilities needed to support comprehensive water management as opposed to the usafructary rights provided by any water withdrawal permitting system of the individual Parties. Naturally, however, the aquisition of real property has associated with it the potential for a "taking" of water rights that may be associated with the land so condemned. This is especially true in jurisdiction, especially those Patties in which appropriative rights have been separated from the land itself.

Providing the Commission with the power of condemnation and eminent domain must be carefully considered. Without such power, the effectiveness of the comprehensive plan might be significantly impaired. However, such power also provides the Commission with the capability of interfering with the inherent sovereignty of the Parties in ways that were unintended. The key to preventing this is in carefully crafting internal laws that address condemnation and invocation of eminent domain.

This provision expressly requires "just compensation" for any taking of property rights. The "just compensation" will, however, depend largely on the individual internal laws of the Parties themselves. In the United States, the recent rulings in regulatory by the Supreme Court have held that a serious impairment of the value of land by a regulation of its use must be compensated, but have noted that the State could diminish the value of a water right by as much as 95% without incurring liability, at least when a system of regulatd riparian rights exist. *Lucas v. South Carolina Coastal Council*, 112 S. Ct. 2886 (1992). *See also* J. Peter Byrne, *The Arguments for the Abolition of the Regulatory Takings Doctrine*, 22 Ecol. L.Q. 89 (1995); Oliver Houck, *Why Do We Protect Endangered Species, and What Does That Say about Whether Restrictions on Private Property to Protect Them Constitute "Takings"?*, 80 Iowa L. Rev. 297 (1995); Joseph Sax, *The Constitution, Property Rights and the Future of Water Law*, 61 U. Colo. L. Rev. 257 (1990).

This may not, however, apply to governments with a system of appropriative rights. Appropriative rights are defined in terms of a specific quantity of water applied to a beneficial use and are much more closely related to propert rights. Such rights to use water are usually vested property that may not be abolished without compensation. *United States v. State Water Resources Control Bd.*, 227 Cal. Rptr. 161 (Cal. App. 1986), review denied.

Cross-references: §2C-2-09 (party or parties); §4C-1-03 (comprehensive water management plan); §4C-1-04 (purpose and objectives of comprehensive water management plan); §4C-1-05 (conditions of comprehensive water management Plan); §4C-1-06 (deviation from comprehensive water management plan); §4C-1-07 (allocation during flood conditions); §4C-1-08 (Allocation during drought conditions); §4C-1-09 (non-impairment of comprehensive water management plan by state action); §4C-2-01 (water allocation, generally); §4C-2-02 (waters not subject to allocation); §4C-2-03 (basin water sources); §4C-2-04 (basin water demands and needs); §4C-2-05 (allocation to equitable and reasonable uses); §4C-2-06 (watershed management); §4C-3-01 (existing rights recognized); §4C-3-02 (flood protection works); §4C-3-03 (minimum flows); §4C-3-04 (withdrawals and diversion; protected areass); §4C-3-05 (water levels protected); §4C-3-06 (augmentation of supply); §4C-3-07 (water quality); §4C-3-08 (underground water; limit on withdrawals); §4C-3-09 (atmospheric water); §4C-3-10 (recreation).

Similar Agreements: *Delaware River Basin Compact*, Pub. L. 87-328, 75 Stat. 688 (1961); *Susquehanna River Basin Compact*, Pub. L. No. 91-575, 84 Stat. 1509 (1970).

§3C-2-10 MEETINGS, HEARINGS AND RECORDS (Optional)

(a) The signatory Parties recognize the importance and necessity of public participation in promoting utilization of the water resource of the _____ River Basin. Consequently, all meetings of the Commission shall be open to the public except with respect to issues of personnel.

(b) The Commission shall conduct at least one public hearing prior to the adoption of the comprehensive plan, water resources program, annual capital and current expense budgets, the letting of any contract for the sale or other disposition by the Commission of hydroelectric energy or water resources to any person, corporation or entity, and in all other cases wherein this Agreement requires a public hearing. Such hearing shall be held upon at least (ten) days public notice given by posting at the offices of the Commission. The Commission shall also provide forthwith for distribution of such notice to the press and by the mailing of a copy thereof to any person who shall request such notices.

(c) The minutes of the Commission shall be a public record open to inspection at its offices during regular business hours.

Commentary. Effective water sharing demands that all stakeholders have information upon which they can rely in order for them to make rational decisions about water use. Without sufficient public participation, the Parties will be unable to maximize the use of the water resource. Although optional, inclusion of this provision is essential to maximum effectiveness of the comprehensive plan that cannot be realized without full disclosure of its administration. Incorporation of this section memorializes the public nature of the enterprise. The US-Canada International Joint Commission has recognized the need for "engaging public support ..." *See* International Joint Commission, Second Biennial Report under the Great Lakes Water Quality Agreement of 1978 to the Governments of the United States and Canada and the Provinces of the Great Lakes Basin (1984). Agenda 21 recognizes the need for the widest cooperation between governmental and non-governmental organizations. However, it is recognized that public notice provisions may not conform to the requirements of the specific jurisdictions involved; jurisdictional requirements control.

Cross-references: §2C-2-06 (equitable and reasonable apportionment); §2C-2-09 (party or parties); §3C-2-01 (general powers and duties); §3C-2-02 (powers and duties reserved to the commissioners); §3C-2-03 (obligations of the commission).

Similar Agreements: *Delaware River Basin Compact*, Pub. L. 87-328, 75 Stat. 688 (1961); *Susquehanna River Basin Compact*, Pub. L. No. 91-575, 84 Stat. 1509 (1970); *Apalachicola-Chattahoochee-Flint River Basin Compact*, O.C.G.A. 12-10-100 (1997); *Alabama-Coosa-Tallapoosa River Basin Compact*, O.C.G.A. 12-10-110 (1997); *Convention on the Protection and Use of Transboundary Watercourses and International Lakes*, 31 I.L.M. 1312 (1992); *The North American Agreement on Environmental Cooperation between the Government of the*

United States of America, the Government of Canada, and the Government of the United Mexican States, 32 I.L.M. 1480 (1993).

§3C-2-11   TORT LIABILITY

The Commission shall be responsible for claims arising out of the negligent acts or omissions of its officers, agents and employees only to the extent and subject to the procedures prescribed by law generally with respect to officers, agents and employees of the government of the Party in which the tortuous conduct occurred.

Commentary. This provision provides accountability by the Commission for actions by its officers, agents and employees. To take effect, similar provisions must exist in the national laws of the respective Parties.

Cross-references: §2C-2-09 (party or parties); §3C-2-01 (general powers and duties); §3C-2-02 (powers and duties reserved to the commissioners); §3C-2-03 (obligations of the commission); §3C-2-09 (condemnation proceedings); §6C-1-01 (good faith implementation).

Similar Agreements: *Delaware River Basin Compact*, Pub. L. 87-328, 75 Stat. 688 (1961); *Susquehanna River Basin Compact*, Pub. L. No. 91-575, 84 Stat. 1509 (1970).

Part 3 Intergovernmental Relations

§3C-3-01   COORDINATION AND COOPERATION

The Commission shall promote and aid the coordination of the activities and programs of governmental and private agencies concerned with water resources administration in the Basin. To this end, but without limitation thereto, the Commission may:

(a) Advise, consult, contract, financially assist, or otherwise cooperate with any and all such agencies;

(b) Employ any other agency or instrumentality of any of the signatory Parties or of any political subdivision thereof, in the design, construction, operation and maintenance of structures, and the installation and management of river control systems, or for any other purpose;

(c) Develop and adopt plans and specifications for particular water resources projects and facilities which so far as consistent with the comprehensive plan incorporate any separate plans of other public and private organizations operating in the Basin, and permit the decentralized administration thereof;

(d) Qualify as a sponsoring agency under any local, national or regional legislation heretofore or hereafter enacted to provide financial or other assistance for the planning, conservation, utilization, development, management or control of water resources.

Commentary. This provision also appears as §1C-1-04 above. Its exact placement is optional. However, the provision is necessary to allow the Commission and its officers to coordinate directly with various political, economic and social entities without needless delay that would be required in the interaction and bureaucratic entanglement in the maze of administrative and legal institutions within the structure of the sovereign Parties that would otherwise be necessary.

Cross-references: §2C-1-02 (consent to jurisdiction); §2C-1-06 (existing agencies); §2C-2-02 (basin); §2C-2-09 (party or parties); §3C-1-05 (commission organization and staffing); §3C-1-06 (rules of procedures); §3C-1-07 (commission administration); §3C-2-01 (general powers and duties); §3C-2-02 (powers and duties reserved to the commissioners); §3C-2-03 (obligations of the commission); §3C-2-06 (referral and review); §3C-2-07 (advisory committees); §3C-3-02 (project costs and evaluation standards); §3C-3-03 (projects of the signatory parties); §3C-3-04 (cooperative services).

Similar Agreements: *Delaware River Basin Compact*, Pub. L. 87-328, 75 Stat. 688 (1961); *Susquehanna River Basin Compact*, Pub. L. No. 91-575, 84 Stat. 1509 (1970); *ASEAN Agreement on the Conservation of Nature and Natural Resources* (1985); *Convention on the Protection and Use of Transboundary Watercourses and International Lakes*, 31 I.L.M. 1312 (1992); *The North American Agreement on Environmental Cooperation between the Government of the United States of America, the Government of Canada, and the Government of the United Mexican States*, 32 I.L.M. 1480 (1993); *Treaty of Peace between the State of Israel and the Hashemite Kingdom of Jordan*, 34 I.L.M. 43 (1994); *Convention on the Law of the Non-Navigational Uses of International Watercourses*, United Nations Document A/51/869 (1998).

§3C-3-02 PROJECT COSTS AND EVALUATION CRITERIA

Consistent with Article 5C, the Commission shall establish uniform criteria and procedures for the evaluation, determination of benefits, and cost allocations of water projects affecting the Basin, and for the determination of project priorities, pursuant to the requirements of the comprehensive plan and its water resources program. The Commission shall develop equitable cost sharing and reimbursement formulas for the signatory Parties including:

(a) Uniform and consistent procedures for the allocation of project costs among purposes included in multiple-purpose programs;

(b) Procedures for allocation of cost sharing between the signatory Parties and with other public and semi-public entities, private organizations and groups according to benefit received;

(c) Establishment and supervision of a system of accounts for reimbursable purposes and directing the payments and charges to be made from such accounts;

Commentary. This provision establishes rational basis for the implementation of comprehensive management of water projects that can survive appropriate economic analysis and cost/benefit scrutiny.

Cross-references: §2C-2-09 (party or parties); §3C-3-03 (projects of the signatory parties); §3C-3-04 (cooperative services); Article 5C (financing).

Similar Agreements: *Delaware River Basin Compact*, Pub. L. 87-328, 75 Stat. 688 (1961); *Susquehanna River Basin Compact*, Pub. L. No. 91-575, 84 Stat. 1509 (1970).

§3C-3-03 PROJECTS OF THE SIGNATORY PARTIES

For the purposes of avoiding conflicts of jurisdiction and giving full effect to the Commission as a regional agency of the signatory Parties, the following rules shall govern projects of the individual Parties and their subdivisions that affect the water resources of the Basin:

(a) The planning of all projects related to powers delegated to the Commission by this Agreement shall be undertaken in consultation with the Commission;

(b) Prior to entering upon the execution of any project authorized by this article, the Commission shall review and consider and integrate into the Comprehensive Water Management Plan insofar as possible all existing rights, plans and programs of the signatory Parties, their political subdivisions, private parties, and water users which are pertinent to such project, and shall hold a public hearing on each proposed project;

(c) No expenditure or commitment shall be made for or on account of the construction, acquisition or operation of any project or facility nor shall it be deemed authorized, unless it shall have first been included by the Commission in the Comprehensive Water Allocation Management Plan;

(d) Each governmental agency otherwise authorized by law to plan, design, construct, operate or maintain any project or facility in or for the Basin shall continue to have, exercise and discharge such authority except as specifically provided by this Article.

Commentary. By this provision, the Parties agree to refrain from independently developing and operating water management facilities that may affect the Comprehensive Water Allocation Management Plan. Without its inclusion, the effectiveness of the Comprehensive Water Management Plan may be weakened.

Cross-references: §2C-2-02 (basin); §2C-2-09 (party or parties); §3C-3-03 (projects of the signatory parties); §3C-3-04 (cooperative services); §4C-1-03 (comprehensive water management plan); §4C-1-04 (purpose and objectives of comprehensive water management plan); §4C-1-05 (conditions of comprehensive water management Plan); §4C-1-06 (deviation from comprehensive water management plan); §4C-1-07 (allocation during flood conditions); §4C-1-08 (Allocation during drought conditions); §4C-1-09 (non-impairment of comprehensive water management plan by state action); Article 5C (financing).

Similar Agreements: *Delaware River Basin Compact*, Pub. L. 87-328, 75 Stat. 688 (1961); *Susquehanna River Basin Compact*, Pub. L. No. 91-575, 84 Stat. 1509 (1970); *ASEAN Agreement on the Conservation of Nature and Natural Resources*, 1985.

§3C-3-04 COOPERATIVE SERVICES

The Commission shall furnish technical services, advice and consultation to authorized agencies of the signatory Parties with respect to the water resources of the Basin, and each of the signatory Parties pledges itself to provide technical and administrative services to the Commission upon request, within the limits of available appropriations and to cooperate generally with the Commission for the purposes of this Agreement, and the cost of such services may be reimbursable whenever the Parties deem appropriate.

Commentary. This provision requires and obliges both the sovereign Parties and the Commission and its officers to provide consultation and services to the other in carrying out the provisions of the Agreement.

Cross-references: §2C-2-09 (party or parties); §3C-3-01 (coordination and cooperation); §3C-3-02 (project costs and evaluation standards); §3C-3-03 (projects of the signatory parties); §4C-1-03 (comprehensive water management plan); §4C-1-04 (purpose and objectives of comprehensive water management plan); §4C-1-05 (conditions of comprehensive water management Plan); §4C-1-06 (deviation from comprehensive water management plan); §4C-1-07 (allocation during flood conditions); §4C-1-08 (Allocation during drought conditions); §4C-1-09 (non-impairment of comprehensive water management plan by state action).

Similar Agreements: *Delaware River Basin Compact*, Pub. L. 87-328, 75 Stat. 688 (1961); *Susquehanna River Basin Compact*, Pub. L. No. 91-575, 84 Stat. 1509 (1970); *Treaty of Peace between the State of Israel and the Hashemite Kingdom of Jordan*, 43 I.L.M. 43 (1994).

ARTICLE 4C

COMPREHENSIVE WATER MANAGEMENT

Part 1 Comprehensive Water Management.

§4C-1-01 JOINT EXERCISE OF SOVEREIGNTY

The water resources of the Basin are subject to the sovereign right and responsibilities of the signatory Parties, and it is the purpose of this Agreement to provide for the joint exercise of such powers of sovereignty over the waters of the _____ River Basin in the common interests of the people of the region.

Commentary. All principal stakeholders directly affected by the shared use must be identified and included in the negotiations. This includes any sovereign government having direct access to surface or underground water. Both governmental and private stakeholders should have a voice in Agreement formulation.

While private water rights holders within the various jurisdictions will not be joined as Parties to the Agreement, it is recommended that the various persons and organizations associated with the various water needs and demands within the basin have a voice in formulating the Agreement. Although the extent of consultation with private groups will depend largely on the political nature of the sovereignties themselves, some recognition of existing rights is required. For instance, in most situations a conflict will exist between demands for water for economic purposes and the needs of environmental protection. Water users frequently need external incentives to accept that some water is reserved for environmental and ecological protection. Unless the governmental entities involved with formulating the Agreement actively seek to include the various interest groups, the effectiveness of the Agreement may be compromised.

Many international agreements reserve the sovereign right of each signatory to exploit its own resources. Some, however, allow for a measure of joint sovereignty but reserve sovereign activities that effect industrial and commercial secrecy or national security. However, recent agreements testify to the trend of joint development of shared water resources and joint development presupposes the joint exercise of sovereignty. *See* J. M. Trolldallen, International Environmental Conflict Resolution; the Role of the United Nations (1992).

Cross-references: §1C-1-01 (general policies); §1C-1-02 (purposes of agreement); §1C-1-03 (objectives of agreement); §2C-1-05 (powers of sovereign parties; withdrawal); §2C-1-06 (existing agencies); §2C-1-07 (limited applicability); §2C-2-02 (basin); §2C-2-03 (comprehensive water management plan); §2C-2-09 (party or parties); §2C-2-11 (waters of the basin); §3C-1-01 (commission created); §3C-1-02 (jurisdiction of the commission); §3C-1-03 (commissioners); §3C-1-02 (jurisdiction of the commission); §3C-1-03 (commissioners); §3C-1-04 (status, immunities and privileges); §3C-1-05 (commission organization and staffing); §3C-1-06 (rules of procedures); §3C-1-07 (commission administration); §3C-2-01 (general powers and duties); §3C-2-02 (powers and duties reserved to the commissioners); §3C-2-03 (obligations of the commission).

Similar Agreements: *Delaware River Basin Compact*, Pub. L. 87-328, 75 Stat. 688 (1961); *Susquehanna River Basin Compact*, Pub. L. No. 91-575, 84 Stat. 1509 (1970); Act regarding Navigation and Economic Cooperation between the States of the Niger Basin, 587 UNTS 9 (1963); Convention and Statute Relating to the Development of the Chad Basin (1974) *reprinted in* Journal officiel de la Republique federale du Cameroon (Yaounde), 4th year. No. 18 (15 Sep. 1964); The Nouakchott Convention Establishing the OMVS (Organisation pour la Mise en Valeur du Fleuve Senegal), 672 UNTS 251 (1974); Convention *on the Law of the Non-Navigational Uses of International Watercourses*, United Nations Document A/51/869 (1998).

§4C-1-02 INTERRELATIONSHIP OF WATER RESOURCES

The water resources of the Basin are functionally interrelated, and the uses of these resources are interdependent. Joint management and coordination of efforts, programs and policies within the Basin is essential to effective and efficient use of the water resource.

Commentary. This provision expands the purpose statement by expressing the intent of the Parties to seek effective water management. Similar provisions are included in Agenda 21 that proposed a program of integrated water resources development and management was necessary.

Cross-references: §1C-1-01 (general policies); §1C-1-02 (purposes of agreement); §1C-1-03 (objectives of agreement); §2C-2-03 (comprehensive water management plan); §4C-1-03 (comprehensive water management plan); §4C-1-04 (purpose and objectives of comprehensive water management plan); §4C-1-05 (conditions of comprehensive water management Plan); §4C-1-06 (deviation from comprehensive water management plan); §4C-1-07 (allocation during flood conditions); §4C-1-08 (Allocation during drought conditions); §4C-1-09 (non-impairment of comprehensive water management plan by state action); §4C-2-01 (water allocation, generally); §4C-2-02 (waters not subject to allocation); §4C-2-03 (basin water sources); §4C-2-04 (basin water demands and needs); §4C-2-05 (allocation to equitable and reasonable uses); §4C-2-06 (watershed management); §4C-3-01 (existing rights recognized); §4C-3-02 (flood protection works); §4C-3-03 (minimum flows); §4C-3-04 (withdrawals and diversion; protected areass); §4C-3-05 (water levels protected); §4C-3-06 (augmentation of supply); §4C-3-07 (water quality); §4C-3-08 (underground water; limit on withdrawals); §4C-3-09 (atmospheric water); §4C-3-10 (recreation).

Similar Agreements: *Delaware River Basin Compact*, Pub. L. 87-328, 75 Stat. 688 (1961); *Susquehanna River Basin Compact*, Pub. L. No. 91-575, 84 Stat. 1509 (1970).

§4C-1-03 COMPREHENSIVE WATER MANAGEMENT PLAN

The Commission shall develop and adopt, and may from time to time review and revise, a Comprehensive Water Management Plan for the immediate and long range development and use of the water resources of the _____ River Basin. The Plan may be as general or specific as the Commission may deem appropriate for efficient and sustainable development of the Basin. At a minimum, the Master Plan shall include:

(a) A water allocation program to allocate the waters of the _____ River Basin with the goal of providing sufficient quantites of quality water resources to satisfy the needs of the Basin for such reasonably foreseeable period as the Commission may determine. This allocation program may supplemented by existing and proposed projects which may be required to satisfy such needs, including all public and private projects affecting the Basin, together with a separate statement of the projects proposed to be undertaken by the Commission during such period; and

(b) A program of water quality management that shall describe, at a minimum, the maximum amounts of designated pollutants allowable in the _____ River and its tributaries;

Commentary. This provision directs the Commission to consider all water-related activities that may affect efficient and sustainable development of the Basin.

The Parties to this Agreement still retain the sovereign right, authority and responsibility to plan and manage their own economic, social and environmental development. Irrespective of whether the planning is the result of State planning or free-market processes, this Agreement suggests that it is the intent of the Parties to provide a baseline of existing and projected water demands. The Commission will integrate the input from the various Parties and develop a plan that equitably optimizes water use among the Parties for economic development within the Basin. Effective basin-wide planning normally requires it be done on a sub-basin, or reach of river, basis. An objective of sustainable development, a significant objective of the *North American Free Trade Agreement*, should be considered. The absence of comprehensive management of a shared resource significantly reduces successful conflict resolution. For instance, the lack of coordinated management has been highlighted as a major hindrance to resolution of the water conflict between India and Pakistan and India and Bangladesh. *See* Clarke, R., Water: The International Crisis (1993).

This section directs the Commission to develop and implement a plan to manage the total water resources of the Basin, subject to certain restrictions, in an effort to maximize the efficiency of water use according to specific goals established by the Agreement. Management of the water allocation can be as detailed as necessary, according to the capabilities and limitations of the Basin. Integration of water quality and quantity is essential. *The UN Convention on the Law of the Non-Navigational Uses of International Watercourses*, United Nations Document A/51/869 (1998), mandates "sustainable development of an international watercourse" and the "rational and optimal utilization, protection and control of the watercourse."

Cross-references: §1C-1-01 (general policies); §1C-1-02 (purposes of agreement); §1C-1-03 (objectives of agreement); §2C-2-02 (basin); §2C-2-03 (comprehensive water management plan); §2C-2-05 (drought); §4C-1-01 (joint exercise of sovereignty); §4C-1-02 (interrelationship of water resources); §4C-1-04 (purpose and objectives of comprehensive water management plan); §4C-1-05 (conditions of comprehensive water management Plan); §4C-1-06 (deviation from comprehensive water management plan); §4C-1-07 (allocation during flood conditions); §4C-1-08 (Allocation during drought conditions); §4C-1-09 (non-impairment of comprehensive water management plan by state action); §4C-2-01 (water allocation, generally); §4C-2-02 (waters not subject to allocation); §4C-2-03 (basin water sources); §4C-2-04 (basin water demands and needs); §4C-2-05 (allocation to equitable and reasonable uses); §4C-2-06 (watershed management); §4C-3-01 (existing rights recognized); §4C-3-02 (flood protection works); §4C-3-03 (minimum flows); §4C-3-04 (withdrawals and diversion; protected areass); §4C-3-05 (water levels protected); §4C-3-06 (augmentation of supply); §4C-3-07 (water quality); §4C-3-08 (underground water; limit on withdrawals); §4C-3-09 (atmospheric water); §4C-3-10 (recreation).

Similar Agreements: *Klamath River Basin Compact*, 71 Stat. 497 (1957); *Delaware River Basin Compact*, Pub. L. 87-328, 75 Stat. 688 (1961); *Susquehanna River Basin Compact*, Pub. L. No. 91-575, 84 Stat. 1509 (1970); Convention *on the Law of the Non-Navigational Uses of International Watercourses*, United Nations Document A/51/869 (1998); Delaware *River Basin Compact*, Pub. L. 87-328, 75 Stat. 688 (1961); Treaty *for Amazonian Cooperation*, 17 ILM 1046 (1978); *Agreement on the Cooperation for the Sustainable Development of the Mekong River Basin*, 34 ILM 864 (1995).

§4C-1-04 PURPOSE AND OBJECTIVES OF COMPREHENSIVE WATER MANAGEMENT PLAN

The purpose of the Comprehensive Water Management Plan is to facilitate maximum utilization of the waters of the Basin to meet both existing and potential needs and uses for sustained economic growth while providing for adequate public health and safety, water-based recreation and environmental protection. A [___]-year planning horizon shall be established. The objectives of the Plan shall include the following:

(a) To identify and assess the sources of water available to the users within the Basin, and to coordinate public and private projects and facilities to augment water availability, which will allow for the optimum planning, development, conservation, utilization, management and control of the water resources of the Basin as described in §4C-2-03 below.

(b) To identify, quantify and value where appropriate all existing and planned water uses within the Basin by sub-Basin and by political jurisdiction as described in §4C-2-04.

(c) To allocate efficiently the available Basin waters according to the needs of the various categories of uses as described in §4C-2-04 and in a manner that ensures equitable and reasonable use of the water.

Commentary. The Parties should agree on a common planning horizon for economic evaluation. In the United States, a 50-year horizon is used by many states for water planning and/or permitting purposes. While the choice of horizon does not in any way limit the duration of the Agreement, equitable partition of the water source requires a common time horizon for the economic evaluation.

Surface and underground water resources are often managed as separate and distinct sources. However, such individualized management does not provide an integrated understanding of either the water sources or their use. Prior to implementing transboundary water use allocation or transfer, participants should conduct a comprehensive water resources assessment. Certain parameters define the framework of the Agreement. The Parties should specify the range of hydrologic events that the Agreement will cover. This means the Parties should clearly establish the quantitative measures of drought and flood conditions and establish the levels when special management for drought or flood conditions arise. The Parties should also define the levels of water quality degradation they are willing to accept as a result of meeting the needs and demands for the water.

The assessment team should identify and quantify the existing and planned water demands in the Basin according to type of use as described below. Optimal water sharing may be accomplished when this is done by sub-Basin or by reach as appropriate. Within each category of use, the team should further classify uses as consumptive or non-consumptive. Demands are classified as consumptive when they remove water from the water source for further use. Their major effect on the watercourse is to reduce the quantity of flow. They may, however, return to the water source a portion of the water used in which case the quality of the receiving waters may be affected. The actual amount of water consumption varies both by type of demand and within the general category of particular demands. It is the efficient allocation of water for consumptive use that presents the greatest challenge to the Commission.

Efficient allocation to equitable and reasonable uses suggests that the various water uses be compared by the economic benefit each use produces wherever possible. Economists contend that the best way to allocate water efficiently is to use market mechanisms. *See* Daniel Hillel, Rivers of Eden (1994). The accepted procedures for measuring benefits and costs are based on the idea that the benefit from making an increment of water available for a particular use is measured by society's willingness to pay for the increment of production resulting from the additional allocation of water. This valuation methodology can easily be applied to water uses that are involved in the "stream of commerce" (e.g., water used to produce agricultural or silvacultural products, water used by the pulp and paper industry, or water supplied by municipalities, to name a few). However, other uses are difficult to measure objectively (e.g., water needed to sustain the aquatic ecology, water needed to sustain hunting and fishing). In some uses of water, adequate methodologies for quantifying benefits have not been developed (e.g., the value of maintaining riparian vegetative buffers). Care must be taken not to neglect these less-easily-valued water uses simply because it is difficult to do so. Additionally, equity is the other broad principle that has guided the allocation of water resources. It is more difficult to articulate a means of achieving equitable allocations than to define rules for maximizing returns.

Although an objective of optimum allocation was not introduced into this model agreement, that objective exists in a number of international agreements. (e.g. *Council on European Communities Third Action Programme on the Environment* (1983); Convention *on the Law of the Non-Navigational Uses of International Watercourses*, United Nations Document A/51/869 (1998).

Cross-references: §1C-1-01 (general policies); §1C-1-02 (purposes of agreement); §1C-1-03 (objectives of agreement); §2C-2-02 (basin); §2C-2-03 (comprehensive water management plan); §2C-2-05 (drought); §4C-1-01 (joint exercise of sovereignty); §4C-1-02 (interrelationship of water resources); §4C-1-03 (comprehensive water management plan); §4C-1-05 (conditions of comprehensive water management Plan); §4C-1-06 (deviation from comprehensive water management plan); §4C-1-07 (allocation during flood conditions); §4C-1-08 (Allocation during drought conditions); §4C-1-09 (non-impairment of comprehensive water management plan by state action); §4C-2-01 (water allocation, generally); §4C-2-02 (waters not subject to allocation); §4C-2-03 (basin water sources); §4C-2-04 (basin water demands and needs); §4C-2-05 (allocation to equitable and reasonable uses); §4C-2-06 (watershed management); §4C-3-01 (existing rights recognized); §4C-3-02 (flood protection works); §4C-3-03 (minimum flows); §4C-3-04 (withdrawals and diversion; protected areass); §4C-3-05 (water levels protected); §4C-3-06 (augmentation of supply); §4C-3-07 (water quality); §4C-3-08 (underground water; limit on withdrawals); §4C-3-09 (atmospheric water); §4C-3-10 (recreation).

Similar Agreements: *Delaware River Basin Compact*, Pub. L. 87-328, 75 Stat. 688 (1961); *Susquehanna River Basin Compact*, Pub. L. No. 91-575, 84 Stat. 1509 (1970); *Agreement on the Cooperation for the Sustainable Development of the Mekong River Basin*, 34 ILM 864 (1995); *Convention on the Law of the Non-Navigational Uses of International Watercourses*, United Nations Document A/51/869 (1998).

§4C-1-05 CONDITIONS OF COMPREHENSIVE WATER MANAGEMENT PLAN (Optional)

(a) **The plan shall allocate waters under all hydrologic conditions, with the exception of floods and droughts, as defined herein.**

(b) **The plan shall include consideration of all public and private projects and facilities which are required, in the judgment of the Commission, for the optimum planning, development, conservation, utilization, management and control of the water resources of the Basin to meet present and future needs; provided that the plan shall include any projects required to conform with any present or future decree or judgment of any court of competent jurisdiction.**

(c) **The basis of the management plan shall be the integration of economic development, maintenance of adequate public health and safety, and environmental protection plans and policies established and submitted by the signatory Parties. However, before the adoption of the plan or any part or revision thereof, the Commission shall consult with water users and interested public bodies and public** utilities and shall consider and give due regard to the findings and recommendations of the various agencies of the signatory Parties and their political subdivisions.

(d) The Commission shall conduct public hearings with respect to the comprehensive management plan prior to the adoption of the plan or any part of the revision thereof.

Commentary. This section provides additional framework for the Comprehensive Plan. Certain provisions are optional however. For instance, the Parties may wish the Comprehensive plan to include the procedures to be used in floods and droughts.

Cross-references: §2C-2-02 (basin); §2C-2-03 (comprehensive water management plan); §2C-2-05 (drought); §2C-2-07 (flood); §2C-2-09 (party or parties); §4C-1-06 (deviation from comprehensive water management plan); §4C-1-07 (allocation during flood conditions); §4C-1-08 (Allocation during drought conditions); §4C-1-09 (non-impairment of comprehensive water management plan by state action); §4C-2-01 (water allocation, generally); §4C-2-02 (waters not subject to allocation); §4C-2-03 (basin water sources); §4C-2-04 (basin water demands and needs); §4C-2-05 (allocation to equitable and reasonable uses); §4C-2-06 (watershed management); §4C-3-01 (existing rights recognized); §4C-3-02 (flood protection works); §4C-3-03 (minimum flows); §4C-3-04 (withdrawals and diversion; protected areass); §4C-3-05 (water levels protected); §4C-3-06 (augmentation of supply); §4C-3-07 (water quality); §4C-3-08 (underground water; limit on withdrawals); §4C-3-09 (atmospheric water); §4C-3-10 (recreation).

Similar Agreements: *Delaware River Basin Compact*, Pub. L. 87-328, 75 Stat. 688 (1961); *Susquehanna River Basin Compact*, Pub. L. No. 91-575, 84 Stat. 1509 (1970); *Agreement on the Cooperation for the Sustainable Development of the Mekong River Basin*, 34 ILM 864 (1995); Convention *on the Law of the Non-Navigational Uses of International Watercourses*, United Nations Document A/51/869 (1998);

§4C-1-06 DEVIATION FROM COMPREHENSIVE WATER MANAGEMENT PLAN

The Commission may authorize deviations from the Comprehensive Water Management Plan may occur during extreme hydrologic conditions, either flood or drought.

Commentary. The Comprehensive Water Management Plan can be crafted to address both flood and drought conditions. However, because the nature and characteristics of extreme hydrologic events are difficult to predict, and because no plan is capable of adequately reacting to all such events, the Plan should have provisions that provide for the unexpected, unplanned for situation.

The issue of climate change has surfaced as a potential impediment to effective long-range policies and management of water resources. At the present time, the accuracy and precision of prediction of the effects is not yet sufficient to provide specific planning capability, the Parties should recognize that the Comprehensive Water Management Plan may require some modifications as predictability of climate changes become more accurate and precise. The American Society of Civil Engineers, in *Policy Statement 360*, recognized the problems associated with water resources planning and climate change:

> Global climate change could pose a potentially serious impact on world-wide water resources, energy production and use, agriculture, forestry, coastal development and resources, flood control and public infrastructure. Such impacts could require modified agricultural practices and could require measures to deal with rising sea levels with associated impacts on estuaries, coastal shoreline land uses and infrastructure. Early participation of the engineering community in the debate and decisions on appropriate solutions is desirable because the practice of engineering deals with the development of cost-effective and environmentally acceptable methods of providing for human needs in relation to the natural environment.

Since the effects of climate change will extend across political boundaries, the Parties should recognize that the Comprehensive Water Management Plan may need to be modified as the continued international research develops more accurate and precise predictability on those changes.

The Intergovernmental Panel on Climate Change of the World Meteorological Organization is specific about the potential problems with lakes, streams and wetlands associated with climate change. *See* IPCC/WMO/UNEP, *Summary for Policymakers: Scientific-Technical Analyses of Impacts, Adaptations and Mitigation of Climate Change - IPCC Working Group II* http://www.ipcc.ch/pub/reports.htm (1994).

> Inland aquatic ecosystems will be influenced by climate change through altered water temperatures, flow regimes and water levels. In lakes and streams, warming would have the greatest biological effects at high latitudes, where biological productivity would increase, and at the low-latitude boundaries of cold- and cool-water species ranges, where extinctions would be greatest. Warming of larger and deeper temperate zone lakes would increase their productivity; although in some shallow lakes and in streams, warming could increase the likelihood of anoxic conditions. Increases in flow variability, particularly the frequency and duration of large floods and droughts, would tend to reduce water quality and biological productivity and habitat in streams. Water-level declines will be most severe in lakes and streams in dry evaporative drainages and in basins with small catchments. The geographical distribution of wetlands is likely to shift with changes in temperature and precipitation. There will be an impact of climate change on greenhouse gas release from non-tidal wetlands, but there is uncertainty regarding the exact effects from site to site.

The IPCC precaution for policy makers presents the problem.

Policymakers will have to decide to what degree they want to take precautionary measures by mitigating greenhouse gas emissions and enhancing the resilience of vulnerable systems by means of adaptation. Uncertainty does not mean that a nation or the world community cannot position itself better to cope with the broad range of possible climate changes or protect against potentially costly future outcomes. Delaying such measures may leave a nation or the world poorly prepared to deal with adverse changes and may increase the possibility of irreversible or very costly consequences. Options for adapting to change or mitigating change that can be justified for other reasons today (e.g., abatement of air and water pollution) and make society more flexible or resilient to anticipated adverse effects of climate change appear particularly desirable.

To restate the problem for the Parties, it is important to recognize the potential problems climate change may cause to the long-range effectiveness of the Comprehensive Water Management Plan. However, predictability of the specific effect of climate change on any specific region are not yet sufficiently accurate or precise to be included in a plan at this time.

Cross-references: §2C-2-03 (comprehensive water management plan); §2C-2-05 (drought); §2C-2-07 (flood); §4C-1-07 (allocation during flood conditions); §4C-1-08 (Allocation during drought conditions); §4C-1-09 (non-impairment of comprehensive water management plan by state action); §4C-2-01 (water allocation, generally); §4C-2-02 (waters not subject to allocation); §4C-3-06 (augmentation of supply); §4C-3-07 (water quality); §4C-3-08 (underground water; limit on withdrawals).

Similar Agreements: *Convention on the Law of the Non-Navigational Uses of International Watercourses*, United Nations Document A/51/869 (1998).

§4C-1-07 ALLOCATION DURING FLOOD CONDITIONS

(a) Retention or release of waters for flood protection operations reasonably needed to prevent loss of life, substantial public disruption, or major property damage shall supersede any other allocation provided in this Agreement. The Commission may plan, design, construct and operate and maintain projects and facilities as it may deem necessary or desirable for flood damage reduction. It shall have power to operate such facilities and to store and release waters on the _____ River Basin, in such manner, at such times, and under such regulations as the Commission may deem appropriate to meet flood conditions as they may arise.

(b) The Commission shall have power to adopt, amend and repeal recommended standards, in the manner provided by this section, relating to the nature and extent of the uses of land in areas subject to flooding by waters of the _____ River and its tributaries. Such standards shall not be deemed to impair or restrict the power of the signatory Parties or their political subdivisions to adopt zoning and other land use regulations not inconsistent therewith.

(c) The Commission may study and determine the nature and extent of the flood plains of the _____ and its tributaries. Upon the basis of such studies, it may establish encroachment lines and delineate the areas subject to flood, including a classification of lands with reference to relative risk of flood and the establishment of standards for flood plain use which will safeguard the public health, safety and property. Prior to the adoption of any standards delineating such area or defining such use, the Commission shall hold public hearings with respect to the substance of such standards. At or before such public hearings the proposed standards shall be available, and all interested persons shall be given an opportunity to be heard thereon at the hearing. Upon the adoption and promulgation of such standards, the Commission may enter into agreements to provide technical and financial aid to any political subdivision for the administration and enforcement of any local land use ordinances or regulations giving effect to such standards.

(d) The Commission shall have power to acquire the fee or any lesser interest in lands and improvements thereon within the area of a flood plain for the purpose of restricting the use of such property so as to minimize the flood hazard, converting property to uses appropriate to flood plain conditions, or preventing unwarranted constrictions that reduce the ability of the river channel to carry flood water.

(e) The Commission may cause lands particularly subject to flood to be posted with flood hazard warnings, and may from time to time cause flood advisory notices to be published and circulated as conditions may warrant.

Commentary. Some preparation can be made to respond to flood conditions. These provisions provide the minimum necessary to prepare for and react to flood conditions.

Cross-references: §2C-2-02 (basin); §2C-2-03 (comprehensive water management plan); §2C-2-07 (flood); §2C-2-09 (party or parties); §4C-1-04 (purpose and objectives of comprehensive water management plan); §4C-1-05 (conditions of comprehensive water management Plan); §4C-1-06 (deviation from comprehensive water management plan); §4C-1-09 (non-impairment of comprehensive water management plan by state action); §4C-2-01 (water allocation, generally); §4C-2-02 (waters not subject to allocation); §4C-3-01 (existing rights recognized); §4C-3-02 (flood protection works); §4C-3-03 (minimum flows); §4C-3-06 (augmentation of supply).

Similar Agreements: *Delaware River Basin Compact*, Pub. L. 87-328, 75 Stat. 688 (1961); *Susquehanna River Basin Compact*, Pub. L. No. 91-575, 84 Stat. 1509 (1970); *Convention on the Law of the Non-Navigational Uses of International Watercourses*, United Nations Document A/51/869 (1998).

§4C-1-08 ALLOCATION UNDER DROUGHT CONDITIONS

The Commission shall, within (____) years of the coming into force of this Agreement, complete the preparation of a Drought Management Plan applicable to the _____ River Basin. The Drought Management Plan shall:

(a) Specify the hydro-meteorological preconditions of a Drought Alert and, thereunder, the conservation measures to be observed by all water users or by individual classes of water users according to a predetermined priority classification in the _____ River Basin;

(b) Specify the hydro-meteorological preconditions of a Drought Emergency and, thereunder, the specific additional measures to be observed by all water users in the _____ River Basin;

(c) Specify the priorities of water demands within the Basin; but *and*

(d) Modify or suspend the conservation and other specific measures provided in the Drought Management Plan in order to meet the specific requirements of the individual drought circumstances.

(e) The Drought Management Plan shall undergo analysis and revision on a five-year cycle, as drought conditions occur, or as circumstances arise.

Commentary. Droughts of any duration, but especially extended droughts, interpose conditions that may require extensive alteration of the allocation prescribed by the Comprehensive Water Allocation Management Plan. During the most extreme droughts, it may be necessary to allocate all water solely for human survival, meaning that water for human consumption be made available along with water for subsistence farming. In other situations, water may be reserved solely for those uses necessary to sustain long-term economic viability. In any event, the Commission should be authorized to deviate from the Plan in a manner that insures equity among the Parties.

Cross-references: §2C-2-03 (comprehensive water management plan); §2C-2-04 (conservation measures); §4C-1-04 (purpose and objectives of comprehensive water management plan); §2C-2-05 (drought); §4C-1-05 (conditions of comprehensive water management Plan); §4C-1-06 (deviation from comprehensive water management plan); §4C-1-09 (non-impairment of comprehensive water management plan by state action); §4C-2-01 (water allocation, generally); §4C-2-02 (waters not subject to allocation); §4C-2-05 (allocation to equitable and reasonable uses); §4C-2-06 (watershed management); §4C-3-04, Withdrawals and Diversion; Protected Areas; §4C-3-05 (water levels protected); §4C-3-06 (augmentation of supply); §4C-3-07 (water quality); §4C-3-08 (underground water; limit on withdrawals); §4C-3-09 (atmospheric water); §4C-3-10 (recreation).

Similar Agreements: *Delaware River Basin Compact*, Pub. L. 87-328, 75 Stat. 688 (1961); *Susquehanna River Basin Compact*, Pub. L. No. 91-575, 84 Stat. 1509 (1970); *Convention on the*

Law of the Non-Navigational Uses of International Watercourses, United Nations Document A/51/869 (1998).

§4C-1-09 NON-IMPAIRMENT OF COMPREHENSIVE WATER MANAGEMENT PLAN BY STATE ACTION

(a) State laws which may allow interbasin transfer of surface water, the transfer of surface water from one part of the Basin to another, or the permitting of water withdrawal rights for private sale, shall be integrated into the Comprehensive Plan by the Commission to the greatest extent possible but only to the degree such integration is consistent with the rights of the State under the terms of the Agreement.

(b) No state laws authorizing interbasin transfer from the _____ River Basin will take effect without the certification by the Commission that such interbasin transfer will not adversely effect the Comprehensive Water Management Plan.

Commentary: Large transfers of water from one water basin to another, or from one part of a single water basin to another, have become common. Often these transfers involve the transfer of water from a rural to an urban area. *See generally* Sax, Abrams, & Thompson, at 262-270; Tarlock, § 10.04(2)(c); Tarlock, Corbridge, & Getches, at 837-39; Trelease & A. Gould, at 78-80; Owen L. Anderson, *Reallocation,* in 2 Water and Water Rights § 16.02(c)(2); Dellapenna, § 7.05(c)(2). Protection for the water basin of origin is now a well-established part of the law of many States, particularly in the western States. *See,* e.g., Ariz. Rev. Stat. Ann. § 45-172; Cal. Water Code § 1215, 10505, 10505.5; Tex. Water Code Ann. § 16.052. Such transfers can be contemplated by the state itself, authorizing the transfer of water resources as part of its economic planning initiatives. Such transfers may also result from the institution of private permitting of water withdrawal for commercial purposes that may include interbasin transfer for private sale. In any event, it is necessary that the agreement acknowledge that the individual Parties may not adopt policies that reduce the available water supply for the Comprehensive Water Management Pla purposes.

Cross-references: §2C-2-02 (basin); §2C-2-08 (interbasin transfer); §4C-1-04 (purpose and objectives of comprehensive water management plan); §2C-2-05 (drought); §4C-1-05 (conditions of comprehensive water management Plan); §4C-1-06 (deviation from comprehensive water management plan); §4C-1-08 (Allocation during drought conditions); §4C-2-01 (water allocation, generally); §4C-2-02 (waters not subject to allocation); §4C-2-03 (basin water sources); §4C-2-04 (basin water demands and needs); §4C-2-05 (allocation to equitable and reasonable uses); §4C-2-06 (watershed management); §4C-3-04, Withdrawals and Diversion; Protected Areas; §4C-3-05 (water levels protected); §4C-3-06 (augmentation of supply); §4C-3-08 (underground water; limit on withdrawals); §4C-3-09 (atmospheric water); §4C-3-10 (recreation).

Similar Agreements: None.

Part 2 Water Allocation

§4C-2-01 WATER ALLOCATION, GENERALLY

(a) The Comprehensive Water Management Plan, updated on an annual basis, shall allocate water resources according to the needs of the various equitable and reasonable uses described in §4C-2-04. The water demands based on application of reasonable conservation standards shall be determined for each sub-Basin or reach, as determined by the Commission, and compared to the water available for that sub-Basin or reach.

(b) If the available water in all sub-Basins exceeds the demand, all reasonable demands, as defined by the Commission, within the Basin shall be fully satisfied.

(c) Allocation shall be first made to existing water rights which have been perfected under applicable law of the individual sovereign Parties prior to the creation of this Agreement. Allocation shall then be made to the needs of environmental protection according to individual environmental policies of the signatory Parties. Allocation thereafter shall be made according to those equitable and reasonable uses which provide maximum economic benefit (whether direct or indirect) to the Parties or which are necessary to enhance specific quality of life objectives of the Parties.

(e) No allocation of waters hereafter made pursuant to this section shall constitute a prior appropriation of the waters of the Basin or confer any superiority of right in respect to the use of those waters, nor shall any such action be deemed to constitute an apportionment of the waters of the Basin among the Parties hereto. This paragraph shall not be deemed to limit or restrict the power of the Commission to enter into covenants with respect to water supply, with a duration not exceeding the life of this Agreement, as it may deem necessary for a benefit or development of the water resources of the Basin.

(f) No signatory Party shall permit any augmentation of flow by release from storage under the direction of the Commission to be diminished by the diversion of any water of the Basin during any period, except in cases where such diversion is duly authorized by this Agreement, or by the Commission pursuant thereto, or by the judgment, order or decree of a court of competent jurisdiction.

(g) Available water in excess of demand will be reserved for use as the Commission may direct.

Commentary. This section describes the fundamental objective function of a water allocation management plan. The UN *Convention on the Law of the Non-Navigational Uses of International Watercourses*, United Nations Document A/51/869 (1998), advises that "(i)n the absence of agreement or custom to the contrary, no use of an international watercourse enjoys inherent priority over other uses." If sufficient water is available to meet all equitable and reasonable uses, all demands will be met and no need for prioritization exists. However, if sufficient water is not available, it is recommended that a priority of allocation be established as follows: vested water rights, minimum instream flow for environmental protection and water quality purposes, followed by allocations that balance demands that provide maximum economic benefit with demands which have a direct influence on quality of life issues but cannot meet the economic threshold. For sustainable development in both transitional and developing countries, there must be an appropriate legal structure and a set of institutions that define property rights. *See* Thomas Sterner, Policy Instruments for a Sustainable Economy, Economic Policies for Sustainable Development (T. Sterner, ed., 1994).

Cross-references: §1C-1-01 (general policies); §1C-1-02 (purposes of agreement); §1C-1-03 (objectives of agreement); §2C-2-02 (basin); §2C-2-03 (comprehensive water management plan); §2C-2-05 (drought); §2C-2-11 (waters of the basin); §4C-1-02 (interrelationship of water resources); §4C-1-03 (comprehensive water management plan); §4C-1-04 (purpose and objectives of comprehensive water management plan); §4C-1-05 (conditions of comprehensive water management Plan); §4C-1-06 (deviation from comprehensive water management plan); §4C-1-07 (allocation during flood conditions); §4C-1-08 (Allocation during drought conditions); §4C-1-09 (non-impairment of comprehensive water management plan by state action); §4C-2-02 (waters not subject to allocation); §4C-2-03 (basin water sources); §4C-2-04 (basin water demands and needs); §4C-2-05 (allocation to equitable and reasonable uses); §4C-2-06 (watershed management); §4C-3-01 (existing rights recognized); §4C-3-02 (flood protection works); §4C-3-03 (minimum flows); §4C-3-04 (withdrawals and diversion; protected areass); §4C-3-05 (water levels protected); §4C-3-06 (augmentation of supply); §4C-3-07 (water quality); §4C-3-08 (underground water; limit on withdrawals); §4C-3-09 (atmospheric water); §4C-3-10 (recreation).

Similar Agreements: *Treaty between the United States of America and Mexico, Utilization of Waters of the Colorado and Tijuana rivers and of the Rio Grande*, 59 Stat. 1219 (1945); Delaware *River Basin Compact*, Pub. L. 87-328, 75 Stat. 688 (1961); *Susquehanna River Basin Compact*, Pub. L. No. 91-575, 84 Stat. 1509 (1970); *Agreement on the Cooperation for the Sustainable Development of the Mekong River Basin*, 34 ILM 864 (1995).

§4C-2-02 WATERS NOT SUBJECT TO ALLOCATION

(a) **Existing water rights which have been perfected under applicable law of the individual sovereign Parties prior to the creation of this Agreement have priority over other water rights and allocations established by this Agreement. Non-perfected water rights have no such priority and will be subject to implementation of this Agreement.**

(b) **The Parties to this Agreement will provide the necessary information to ensure the Commission can allocate sufficient water within the Comprehensive**

Water Allocation Management Plan to meet the needs of vested water rights holders.

Commentary. This section acknowledges that certain water rights have been vested by internal water laws and that these vested water rights are acknowledged and included within the Commission's planning. Initial identification of vested water rights, and knowledge of laws which may vest such rights in the future is necessary to prevent extremely complex water problems such as those experienced by Los Angeles as Arizona and Nevada exercise their legal rights to the waters of the Colorado River. It should be noted, however, that if all waters within one or more of the Parties are already vested, either the Agreement is meaningless or the internal laws of the particular Parties must be changed.

Cross-references: §2C-2-09 (party or parties); §4C-2-01 (water allocation, generally); §4C-2-03 (basin water sources); §4C-2-04 (basin water demands and needs); §4C-2-05 (allocation to equitable and reasonable uses); §4C-2-06 (watershed management); §4C-3-01 (existing rights recognized); §4C-3-02 (flood protection works); §4C-3-03 (minimum flows); §4C-3-04 (withdrawals and diversion; protected areass); §4C-3-05 (water levels protected); §4C-3-06 (augmentation of supply); §4C-3-07 (water quality); §4C-3-08 (underground water; limit on withdrawals); §4C-3-09 (atmospheric water); §4C-3-10 (recreation).

Similar Agreements: *Delaware River Basin Compact*, Pub. L. 87-328, 75 Stat. 688 (1961); *Susquehanna River Basin Compact*, Pub. L. No. 91-575, 84 Stat. 1509 (1970).

§4C-2-03 ASSESSMENT AND ENHANCEMENT OF BASIN WATER SOURCES

(a) The Comprehensive Water Allocation Management Plan will identify and examine the factors that influence the availability of water resources in the Basins through an assessment of climatology, physiography, geology and existing underground and surface water resources, including reservoirs, and the interaction between underground and surface water resources. The assessment will determine the existing and potential future availability and quality of underground and surface water resources within the Basins, the gains and losses of potential out-of-Basin transfers and the costs associated with making such waters available to users.

(b) The Commission shall have the power to develop and implement plans and projects for augmenting the supply of water, both geographically and temporally, for equitable and reasonable uses of water in the Basin. To this end, without limitation thereto, it may provide for, construct, acquire, operate and maintain dams, reservoirs and other facilities for the utilization of surface and underground water resources, and all related structures, appurtenances and equipment in the river Basin and its tributaries and at such off-river sites as it may find appropriate, and may regulate and control use thereof.

(c) The Commission may contract for the transfer of water from outside the Basin when and if such transfer is made for equitable and reasonable purposes.

(d) The Comprehensive Water Management Plan shall include a water availability model that will allow the water availability in specific reaches or sub-Basins to be determined according to planned or programmed water usage.

Commentary. Proper water resources assessment is essential. *See Council of European Communities Second Action Program on the Environment* (1977). An assessment of surface waters must include the existing quantity and quality of the available surface water as well as the potential losses resulting from man-made changes to the physiographic and climatic factors within the river basin. The assessment must also consider transfers into and out of the river Basin under study. Water demands by a variety of users can dramatically alter both the quantity and quality, as well the temporal and spatial availability of water. Physiographic and climatic factors determine the rate and distribution of runoff within a Basin.

In the underground water assessment, the Parties should develop an ongoing aquifer assessment program. Periodically, on a seasonal or annual basis, the Parties should estimate the total water balance and the "safe yield" for the aquifers. This would entail, in part: (a) identifying the location and quantifying the rate of natural recharge to the aquifer; (b) determining the quantity and rate of diversion, the consumptive use of water, and the rate of natural discharge from the aquifer; (c) estimating changes in underground water storage or flow due to withdrawals; and (d) quantify the relationships between underground water recharge, water table elevations, and aquifer discharges.

Cross-references: §2C-2-02 (basin); §2C-2-03 (comprehensive water management plan); §2C-2-05 (drought); §2C-2-10 (underground water); §4C-2-01 (water allocation, generally); §4C-2-02 (waters not subject to allocation); §4C-2-06 (watershed management); §4C-3-01 (existing rights recognized); §4C-3-02 (flood protection works); §4C-3-03 (minimum flows); §4C-3-04 (withdrawals and diversion; protected areass); §4C-3-05 (water levels protected); §4C-3-06 (augmentation of supply); §4C-3-08 (underground water; limit on withdrawals); §4C-3-09 (atmospheric water); §4C-3-10 (recreation).

Similar Agreements: *Delaware River Basin Compact*, Pub. L. 87-328, 75 Stat. 688 (1961); *Susquehanna River Basin Compact*, Pub. L. No. 91-575, 84 Stat. 1509 (1970); *Treaty between the United States of America and Mexico, Utilization of Waters of the Colorado and Tijuana rivers and of the Rio Grande*, 59 Stat. 1219 (1945); *Convention on the Law of the Non-Navigational Uses of International Watercourses*, United Nations Document A/51/869 (1998).

§4C-2-04 ASSESSMENT AND CLASSIFICATION OF BASIN WATER DEMANDS AND NEEDS

Existing and planned water uses shall be identified and quantified according to the categories described below by sub-Basin or by reach of the _____ River. Within each category, uses will be furthered classified as consumptive or non-consumptive.

Water use shall be determined based on existing and projected usage for the following classes of use:

(a) **Agricultural water uses.**
(b) **Public water supply.**
(c) **Environmental water Needs;**
(d) **Hydropower demands.**
(e) **Industrial demands.**
(f) **Navigation water Needs;**
(g) **Recreation and scenic beauty water requirements.**
(h) **Thermoelectric and nuclear power needs;**
(i) **Waste assimilation.**
(j) **Out-of-Basin transfers.**

Commentary. These categories have been used in effective water resources management in the United States for many years. Other categorizing methods may be used, however, at the discretion of the Parties.

Agriculture is the most significant consumer of water. Approximately 85% of total global water use is in irrigation. *See* Clarke, R., Water: The International Crisis (1993). In the United States, agricultural use accounts for over 42% of all freshwater withdrawals. Within the United States, approximately 80% of the agricultural use can be considered consumptive. *See* U.S. Geologic Survey National Water Summary (1987). Approximately 90% of western water consumption is for irrigation. *See* Allen V. Kneese, *Economics and Water Resources,* Water Resource Administration in Water Resources Administration in the United States (M. Reuss, ed., 1993). Accurate projection of water needed for agricultural programs in developing countries is especially critical. Crop irrigation should be quantified to reflect the quantity of water required using existing irrigation techniques and the quantity required using the most efficient techniques, as determined by the Commission.

Public supply includes water provided by utilities for public functions, including domestic and commercial uses. Domestic uses include water for normal household purposes such as drinking, food preparation, bathing, washing clothes and watering lawns and gardens. Commercial uses include water for the service industry and for retail facilities. It does not include water for industry. That portion of the water withdrawn for public supply and not returned to the water source from which it is drawn is consumptive. Public supply, also know as domestic/commercial use, consumes approximately 20% of the water removed from the water sources. Quantification usually may be determined according to governmental water supply agency records. A minimum usage per person, often set at 100 gallons per day (379 liters per day), may be established as a minimum allocation for public supply use.

Environmental needs, defined as the need for water by natural biological systems, includes environmental protection uses but excludes recreational and scenic beauty uses. Estimation of the need is unique because determining the water needed for ecological sustainability goes beyond simple analysis of the quantity and quality of water needed in the short-term to prevent significant harm to the environment. Rather, environmental demand should include analysis of the value society places on specific aspects of the environment and the tradeoffs it is willing to make. Unlike the other water demands, environmental demand must also consider the long-term affects of not meeting the water needs of natural systems. The political process should apportion environmental sustainability needs by imposing minimum streamflows at selected stations. Special consideration will be given to the habitat requirements of endangered species. Sufficient instream flow of sufficient quality shall be set aside for these purposes. These minimum flows should be established in the economic development and environmental protection plans established and submitted by the signatory Parties. Such minimum streamflows should be sufficient to ensure that other uses or development do not degrade existing fish and game resources. Special consideration should be given to the habitat requirements of endangered species. Instream flow of sufficient quality should be set aside for these purposes. Consideration of quantity and quality, including temperature, are important. Maintenance and restoration of natural environment is stressed in *Council on European Communities Third Action Programme on the Environment*, 1983. It is recommended that the demand be apportioned according to the minimum streamflow at selected stations as determined by the Comprehensive Water Allocation Management Plan. However, as it has been learned in Great Britain, it must be recognized that effective application of the minimum flows concept requires reliable long-term flow data to determine flows accurately, as well as analysis based on the scientific method that determines the flows necessary to maintain habitats. (W. Howarth, The Law of the National Rivers Authority (1990).

Hydropower uses can be identified and quantified according to the existing hydroelectric generating capacity and the flow necessary to support that capacity. The use shall be considered non-consumptive. It usually may be quantified in terms of stable reservoir levels, river flow regimes and water quality. For instance, hydropower demand may be quantified according to the seasonal minimum flows necessary to maintain the necessary generator head and navigation demand according to the flow requirements necessary to maintain the appropriate channel depths. Although non-consumptive with regard to water quantity, hydropower use should be classified as consumptive within the particular sub-Basin or reach since the thermal effects cause significant water quality concerns for certain other uses.

Industrial demand includes process water for industrial, manufacturing and mining purposes. On average, approximately 15% of the water withdrawn is consumed. Industrial demand does not include cooling water and steam for fossil fuel and nuclear power plants since the use of cooling water may be considered nonconsumptive.

Navigation uses, as non-consumptive and an instream flow demand, shall be quantified according to the flow requirements necessary to maintain the appropriate channel depth, and, therefore, necessary minimum river flows.

Recreation and scenic beauty water use shall be quantified according to the need for stable reservoir levels, river flow regimes and water quality.

Thermoelectric and nuclear power demand, when withdrawn for evaporative cooling purposes, consumes less than 3% of the water removed from the water source. Under these conditions, waters used for evaporative cooling purposes are non-consumptive with regard to water quantity. However, this demand has thermal effects on the returned water that may cause significant water quality concerns for certain other uses. Water withdrawn for closed water cooling systems may be considered as 100% consumptive.

Waste assimilation is an important use of water. In the United States, current U.S. Environmental Protection Agency regulations prohibit waste assimilation as a valid category of water use. However, the reality around the world is that waste assimilation is arguably a significant purpose of surface waters. It is suggested that the demand be quantified according to the minimum instream flows necessary for waste assimilation at various levels of treatment (e.g., primary, secondary and tertiary treatment). Such minimum flow quantities should include the water necessary to adequately nourish wetland and riparian buffers whose ability to filter nitrates and phosphorous can clearly assist in waste assimilation.

Out-of-Basin transfers should be recognized as a demand on the waters available to users in the Basin. Both existing and potential out-of-Basin demands should be determined. This demand may significantly reduce the future availability of water due to either existing transfers already recognized in internal water laws or in those cases wherein national water laws allow private transfers under a free market approach to water allocation. As vested under national water laws, out-of-Basin transfers should be specified in national economic development plans provided to the Commission.

Cross-references: §2C-2-03 (comprehensive water management plan); §2C-2-05 (drought); §2C-2-10 (underground water); §4C-2-01 (water allocation, generally); §4C-2-02 (waters not subject to allocation); §4C-2-04 (basin water demands and needs); §4C-2-05 (allocation to equitable and reasonable uses); §4C-2-06 (watershed management); §4C-3-01 (existing rights recognized); §4C-3-02 (flood protection works); §4C-3-03 (minimum flows); §4C-3-04 (withdrawals and diversion; protected areass); §4C-3-05 (water levels protected); §4C-3-06 (augmentation of supply); §4C-3-07 (water quality); §4C-3-08 (underground water; limit on withdrawals); §4C-3-09 (atmospheric water); §4C-3-10 (recreation).

Similar Agreements: *Delaware River Basin Compact*, Pub. L. 87-328, 75 Stat. 688 (1961); *Susquehanna River Basin Compact*, Pub. L. No. 91-575, 84 Stat. 1509 (1970); *Convention on the Law of the Non-Navigational Uses of International Watercourses*, United Nations Document A/51/869 (1998).

§4C-2-05   CRITERIA FOR WATER ALLOCATION (Optional)

(a) Allocation of the water among the various equitable and reasonable uses shall be made on the basis of an analysis of the benefit provided to the user as compared to the detriment suffered by other uses that do not receive an allocation. The benefit evaluation shall include at a minimum (1) economic benefit, (2) environmental protection needs and minimum flows as determined by state laws, and (3) water uses which the Parties determine are required to support social and public policies and purposes.

(b) The Commission shall develop and publish a standard methodology for benefit evaluation of each category, and sub-category where applicable, of water demand to be used by all Parties to this Agreement.

Commentary. The essence of current international law establishes "equitable and reasonable utilization" as the basis for water policy. To implement the policy, however, standards must be established that describe the conditions under which "equitable and reasonable utilization" occur.

The definition of the value of water has been one of the important sources of controversy with respect to water policy. Natural resources economists tend to focus on water as a commodity and argue that water should be allocated to those uses that bring the highest economic return. Governmental subsidies are discouraged as causing distortions of price. Others support subsidies as important to issues of national interest, equity, prior commitments and regional and local development. Current discussion concerning valuation of water as a commodity has met resistance from some economists and political scientists who argue that some persons in society, specifically the poor, are materially disadvantaged in the bargaining process and thus may be willing to give up their water rights for a short-term, and potentially damaging, advantage. They claim water allocation should be a matter of environmental ethics and justice. *See* Dean E. Mann, *Political Science: The Past and Future of Water Resources Policy and Management*, in Water Resources Administration in the United States (M. Reuss, ed., 1993). *See also* Kenneth D. Frederick, *The Economics of Risk in Water Resources Planning* in Water Resources Administration in the United States (M. Reuss, ed., 1993). In the United States, the claim has been made that it is not possible to include relative economic benefit as an allocation criteria because the "relative value of water uses can be determined only through an interactive process of voluntary exchange among thousands of water users." (*See* Phyllis Park Saarinen, and Gary D. Lynn, *Getting The Most Valuable Water Supply Pie: Economic Efficiency in Florida's Reasonable-Beneficial Use Standard*, J. of Land Use & Environmental Law, Vol. 8, No. 2, Summer 1993 Supplement).

Those comments notwithstanding, economic evaluation of allocation is mandatory in some form. The benefit evaluation techniques described below are relevant to free-market economies. Where such economies do not exist, other means should be undertaken to value the different categories of water use.

The economic assessment should include both intra- and inter-Party economic return for existing and potential water use as well as the potential for commercial transfers as a means for resolving conflicts between the the Parties. The allocation process should involve tradeoffs between the various uses, in terms of their individual benefit-cost ratio, as well as tradeoffs with the minimum flows established for environmental sustainability. To begin the assessment, the Parties should estimate the intrastate demand water that provides optimal economic return for each Party over a specific planning horizon based on a standard rate of return. The Parties may then make tradeoffs among themselves for regional allocation of the water resource.

Using the concept of marginal willingness to pay is of limited effectiveness in evaluating environmental sustainability demand. While many economists have proposed that a marginal willingness to pay may be estimated using techniques similar to those used for recreational demand, such techniques may be grossly inexact since the consumer in this case can be classified as the society at large. Contingent valuation has become the most widely used approach to value pubic goods. However, there are a number of problems associated with a contingent valuation study, such as: a vague or unclear description of the good, lack of key information about context and substitutes, lack of attention to ways respondents could misperceive the good, implausible or overly hypothetical scenarios (particularly with respect to provision of the good and payment for it), willingness to pay responses which cannot be predicted by the available covariates, inadequate sampling procedures, poor response rates, and sample sizes which are too small for the purpose for which they were intended. *See* Robert Cameron Mitchell and Richard T. Carson, *Current issues in the design, administration, and analysis of contingent valuation surveys*, Current Issues in Environmental Economics, (Johansson, Kristrom and Male, eds., 1995).

The direct approach to the estimation of willingness-to-pay is to ask individuals how much they are willing to pay for a risk reduction. *See* Magnus Johannsson, Per-Olov Johansson, Bengst Jonsson and Tore Soderqvist, *Valuing changes in health: theoretical and empirical issues*, Current Issues in Environmental Economics, (Johansson, Kristrom and Male, eds., 1995). It has been suggested that the damages could be approximated by the cost for reaching an environmental goal. *See* Partha Dasgupta, Bengt Kristrom, and Karl Goran Maler, *Current issues in resource accounting*, Current Issues in Environmental Economics, (Johansson, Kristrom and Male, eds., 1995). Thus, if the officially stated goal is to reduce sulphur-emissions by thirty per cent, then one can approximate current sulphur-damages by the cost of reaching this goal. The willingness to sacrifice some degree of environmental sustainability relates directly to the political will of the national and/or international community. We therefore recommend allocating environmental sustainability demand according to an established minimum flow determined as discussed above.

Certain water uses exist which are associated with quality-of-life issues or which have social and/or civic purposes that cannot be appropriately quantified in a valid benefit evaluation. In a manner similar to minimum flows established for environmental protection purposes, these equitable and reasonable uses must be established by the Parties within their internal political process and vested in agreement with the other Parties.

Cross-references: §1C-1-01 (general policies); §1C-1-02 (purposes of agreement); §1C-1-03 (objectives of agreement); §2C-2-03 (comprehensive water management plan); §2C-2-05 (drought); §2C-2-09 (party or parties); §4C-1-03 (comprehensive water management plan); §4C-1-04 (purpose and objectives of comprehensive water management plan); §4C-1-05 (conditions of comprehensive water management Plan); §4C-1-06 (deviation from comprehensive water management plan); §4C-1-07 (allocation during flood conditions); §4C-1-08 (Allocation during drought conditions); §4C-1-09 (non-impairment of comprehensive water management plan by state action); §4C-2-01 (water allocation, generally); §4C-2-02 (waters not subject to allocation); §4C-2-03 (basin water sources); §4C-2-04 (basin water demands and needs); §4C-2-06 (watershed management); §4C-3-01 (existing rights recognized); §4C-3-02 (flood protection works); §4C-3-03 (minimum flows); §4C-3-04 (withdrawals and diversion; protected areass); §4C-3-05 (water levels protected); §4C-3-06 (augmentation of supply); §4C-3-07 (water quality); §4C-3-08 (underground water; limit on withdrawals); §4C-3-09 (atmospheric water); §4C-3-10 (recreation).

Similar Agreements: *Klamath River Basin Compact*, 71 Stat. 497 (1957); *Delaware River Basin Compact*, Pub. L. 87-328, 75 Stat. 688 (1961); *Susquehanna River Basin Compact*, Pub. L. No. 91-575, 84 Stat. 1509 (1970); *Agreement on the Cooperation for the Sustainable Development of the Mekong River Basin*, 34 ILM 864 (1995); Convention *on the Law of the Non-Navigational Uses of International Watercourses*, United Nations Document A/51/869 (1998).

§4C-2-06   WATERSHED MANAGEMENT (Optional)

(a) The Commission shall promote sound practices of watershed management in the Basin, including projects and facilities to inhibit uncontrolled

runoff and prevent soil erosion. Natural solutions, such as riparian vegetated buffers, will be proposed whenever possible.

(b) The Commission may acquire, sponsor or operate facilities and projects to encourage soil conservation prevent and control erosion, and to promote land reclamation and sound forestry practices.

(c) The Commission may acquire, sponsor or operate projects and facilities for the maintenance and improvement of fish and wildlife habitats related to the water resources of the Basin.

(d) The Commission shall not operate any such project or facility unless it has first found and determined that no other suitable unit or agency of government is available to operate the same upon reasonable conditions, in accordance with the intent and purpose expressed in Section 2C-1-06 of this Agreement.

Commentary. This section reinforces the philosophy that efficient and effective allocation can be best established by use of watershed management techniques.

Cross-references: §1C-1-01 (general policies); §1C-1-02 (purposes of agreement); §1C-1-03 (objectives of agreement); §2C-2-02 (basin); §4C-1-03 (comprehensive water management plan); §2C-2-05 (drought); §4C-1-04 (purpose and objectives of comprehensive water management plan); §4C-1-05 (conditions of comprehensive water management Plan); §4C-1-06 (deviation from comprehensive water management plan); §4C-1-07 (allocation during flood conditions); §4C-1-08 (Allocation during drought conditions); §4C-1-09 (non-impairment of comprehensive water management plan by state action); §4C-2-01 (water allocation, generally); §4C-2-02 (waters not subject to allocation); §4C-2-03 (basin water sources); §4C-2-04 (basin water demands and needs); §4C-2-05 (allocation to equitable and reasonable uses); ; §4C-3-01 (existing rights recognized); §4C-3-02 (flood protection works); §4C-3-03 (minimum flows); §4C-3-04 (withdrawals and diversion; protected areass); §4C-3-05 (water levels protected); §4C-3-06 (augmentation of supply); §4C-3-07 (water quality); §4C-3-08 (underground water; limit on withdrawals); §4C-3-09 (atmospheric water); §4C-3-10 (recreation).

Similar Agreements: Convention on Wetlands of International Importance, especially as Waterfowl Habitat, International Environmental Law, Multilateral Agreements, 971:09 (1971); Rio Declaration on Environment and Development, 31 I.L.M. 1529 (1992); Agreement on the Protection of the (River) Meuse, 34 ILM 854 (1995); *Agreement on the Cooperation for the Sustainable Development of the Mekong River Basin*, 34 ILM 864 (1995).

Part 3 General Provisions

§4C-3-01 EXISTING RIGHTS RECOGNIZED

Rights to the use of the Waters of the Basin, whether instream use or based on direct diversion or storage, are hereby recognized as of the date of this agreement to the extent these rights are valid under the law of the Party in which the use is made, and shall remain unimpaired hereby. These rights, together with the additional allocations made under §4C-2-05, are agreed to be an equitable and reasonable utilization of the Waters of the Basin among the Parties.

Commentary: This provision is recognition of the inherent sovereignty of the individual Parties and the inherent rights and privileges that the individual Party may have granted for use of water within its borders. The Agreement acknowledges the continuation of previously granted property rights. The Parties do not cede to the Comprehensive Water Management Plan the right to withdraw those water rights existing under state law prior to the Agreement.

Cross-references: §2C-2-02 (basin); §2C-2-03 (comprehensive water management plan); §2C-2-06 (equitable and reasonable apportionment); §2C-2-09 (party or parties); §2C-2-11 (waters of the basin); §4C-1-01 (joint exercise of sovereignty); §4C-1-02 (interrelationship of water resources); §4C-1-03 (comprehensive water management plan); §2C-2-05 (drought); §4C-1-04 (purpose and objectives of comprehensive water management plan); §4C-1-05 (conditions of comprehensive water management Plan); §4C-1-06 (deviation from comprehensive water management plan); §4C-1-07 (allocation during flood conditions); §4C-1-08 (Allocation during drought conditions); §4C-1-09 (non-impairment of comprehensive water management plan by state action); §4C-2-01 (water allocation, generally); §4C-2-02 (waters not subject to allocation); §4C-2-03 (basin water sources); §4C-2-04 (basin water demands and needs); §4C-2-05 (allocation to equitable and reasonable uses); §4C-2-06 (watershed management); §4C-3-01 (existing rights recognized); §4C-3-02 (flood protection works); §4C-3-03 (minimum flows); §4C-3-04 (withdrawals and diversion; protected areass); §4C-3-05 (water levels protected); §4C-3-06 (augmentation of supply); §4C-3-07 (water quality); §4C-3-08 (underground water; limit on withdrawals); §4C-3-09 (atmospheric water); §4C-3-10 (recreation).

Similar Agreements: *Red River Compact*, 94 Stat. 3305 (1978); *Sabine River Compact*, 68 Stat. 690 (1953); amended 76 Stat. 34 (1962); 91 Stat. 281 (1977); and 106 Stat. 4661 (1992).

§4C-3-02 FLOOD PROTECTION WORKS

(a) As a general concept, the use of the channels of the waters of the _____ River Basin for the discharge of flood or other excess waters shall be free and not subject to limitation by any Party, and no Party shall have any claim against the other in respect of any damage caused by such use. However, each of the signatory Parties declares its intention to manage flood control programs and activities in such manner, consistent with the normal operations of its hydraulic systems, as to avoid, as far as feasible, material damage in the territory of the other.

(b) The signatory Parties agree to furnish the Commission with complete documentation of existing flood protection programs and works. The Commission shall analyze the documentation to determine the potential flood damages that may therefrom arise and consult with the signatory Parties concerning the findings of such analysis.

Commentary: Flood control policies and works can have a dramatic effect on the timing and elevation of water levels and thus may become a major contentious issue between the Parties. The issue often will eclipse the boundaries of "equitable and reasonable utilization" and should be addressed as an individual area of coordination. This provision recognizes the sovereign right of each Party to make efforts to safeguard its people and economic assets from flood damages but establishes an avenue for the sharing of data on flood control efforts as well as an independent analysis of the effects of those efforts on other Parties.

Cross-references: §1C-1-01 (general policies); §1C-1-02 (purposes of agreement); §1C-1-03 (objectives of agreement); §2C-2-02 (basin); §2C-2-03 (comprehensive water management plan); §2C-2-07 (flood); §2C-2-09 (party or parties); §3C-3-03 (projects of the signatory parties); §3C-3-04 (cooperative services); §4C-1-01 (joint exercise of sovereignty); §4C-1-02 (interrelationship of water resources); §4C-1-03 (comprehensive water management plan); §4C-1-04 (purpose and objectives of comprehensive water management plan); §4C-1-05 (conditions of comprehensive water management Plan); §4C-1-06 (deviation from comprehensive water management plan); §4C-1-07 (allocation during flood conditions).

Similar Agreements: *Utilization of Waters of the Colorado and Tijuana Rivers and of the Rio Grands, Treaty between the United States of America and Mexico,* 59 Stat. 1219 (1945); *Agreement Between the People's Republic of Bulgaria and the Republic of Turkey Concerning Co-operation in the Use of the Waters of Rivers Flowing through the Territory of Both Countries,* UNTS, Vol. 807, 117 (1968).

§4C-3-03 MINIMUM FLOWS

Alternative 1:

Any requirements by either Party for water to maintain minimum flows or for instream uses shall be satisfied from the allocation provided to that Party.

Alternative 2:

Any allocation formula shall apply after sufficient allowance is made to satisfy requirements for maintaining minimum flow or for instream use.

Commentary: For ecological reasons and for wastewater discharge purposes, a minimum flow must be maintained in all surface waters. This protected minimum flow or level for each water source, normally established by State law, is not subject to allocation for other purposes. In the transboundary context, conflict may occur because the Parties may differ on the quantification of this protected level.

The alternatives presented in this provision provide a model for two different eventualities. Alternative 1 is provided for those cases in which the Parties cannot agree on the specific minimum flow or level to be maintained in the watercourse. Alternative 2 provides a model for the case in which an agreed-upon minimum flow is set in the Agreement.

Cross-references: §1C-1-01 (general policies); §1C-1-02 (purposes of agreement); §1C-1-03 (objectives of agreement); §2C-2-02 (basin); §2C-2-03 (comprehensive water management plan); §2C-2-05 (drought); §2C-2-09 (party or parties); §2C-2-10 (underground water); §4C-1-02 (interrelationship of water resources); §4C-1-03 (comprehensive water management plan); §4C-1-04 (purpose and objectives of comprehensive water management plan); §4C-1-05 (conditions of comprehensive water management Plan); §4C-1-06 (deviation from comprehensive water management plan); §4C-1-08 (Allocation during drought conditions); §4C-1-09 (non-impairment of comprehensive water management plan by state action); §4C-2-01 (water allocation, generally); §4C-2-02 (waters not subject to allocation); §4C-2-03 (basin water sources); §4C-2-04 (basin water demands and needs).

Similar Agreements: *Red River Compact*, 94 Stat. 3305 (1978); *Sabine River Compact*, 68 Stat. 690 (1953); amended 76 Stat. 34 (1962); 91 Stat. 281 (1977); and 106 Stat. 4661 (1992).

§4C-3-04 WITHDRAWALS AND DIVERSIONS; PROTECTED AREAS (Optional)

(a) **The Commission may regulate and control withdrawals and diversions from surface waters and underground waters of the Basin, as provided by this Article. The Commission may enter into agreements with the signatory Parties relating to the exercise of such power or regulation or control and may delegate to any of them such powers of the Commission as it may deem necessary or desirable.**

(b) **(Optional) The Commission may from time to time, after public hearing upon due notice, determine and delineate such areas within the Basin wherein the demands upon supply made by water users have developed or threaten to develop to such a degree as to create a water shortage or to impair or conflict with the requirements or effectuation of the Comprehensive Water Management Plan, and any such areas may be designated as protected areas. The Commission, whenever it determines that such shortage no longer exists, shall terminate the protected status of such area and shall give public notice of such termination.**

(c) **(Optional) In any protected areas so determined and delineated, no person, firm, corporation or other entity shall divert or withdraw water for domestic, municipal, agricultural or industrial uses in excess of such quantities as the Commission may prescribe by general regulation, except (i) pursuant to a permit granted under this article, or (ii) pursuant to a permit or approval heretofore granted under the laws of any of the signatory Parties.**

(d) Permits shall be granted, modified or denied as the case may be so as to avoid such depletion of the natural streamflows and underground waters in the protected area or in an emergency area as will adversely affect the Comprehensive Water Management Plan or the just and equitable interests and rights of other lawful users of the same source, giving due regard to the need to balance and reconcile alternative and conflicting uses in the event of an actual or threatened shortage of water of the quality required.

(e) Each Party shall provide for the maintenance and preservation of such records of authorized diversions and withdrawals and the annual volume thereof as the Commission shall prescribe. Such records and supplementary reports shall be furnished to the Commission at its request.

(f) Whenever the Commission finds it necessary or desirable to exercise the powers conferred by this section, any diversion or withdrawal permits authorized or issued under the laws of any of the signatory Parties shall be superseded to the extent of any conflict with the control and regulation exercised by the Commission.

Commentary. Optimum management of water resources requires that a central authority regulate withdrawals and diversions from surface waters and underground waters of the Basin. However, this optional article has significant impact on the sovereign integrity of the individual state's control over internal affairs.

Cross-references: §2C-1-05 (powers of sovereign parties; withdrawal); §2C-2-03 (comprehensive water management plan); §2C-2-05 (drought); §2C-2-09 (party or parties); §2C-2-10 (underground water); §2C-2-11 (waters of the basin); §3C-2-01 (general powers and duties); §3C-2-02 (powers and duties reserved to the commissioners); §3C-2-03 (obligations of the commission); §3C-2-04 (regulations; enforcement); §3C-2-05 (prohibited activities); §4C-1-01 (joint exercise of sovereignty); §4C-1-02 (interrelationship of water resources); §4C-1-03 (comprehensive water management plan); §4C-1-04 (purpose and objectives of comprehensive water management plan); §4C-1-05 (conditions of comprehensive water management Plan§4C-1-09 (non-impairment of comprehensive water management plan by state action); §4C-2-04 (basin water demands and needs); §4C-2-05 (allocation to equitable and reasonable uses); §4C-2-06 (watershed management).

Similar Agreements: *Delaware River Basin Compact*, Pub. L. 87-328, 75 Stat. 688 (1961); *Susquehanna River Basin Compact*, Pub. L. No. 91-575, 84 Stat. 1509 (1970); *Treaty between the United States and Great Britain relating to Boundary Waters*, 36 Stat. 2451 (1909); ASEAN Agreement on the Conservation of Nature and Natural Resources (1985).

§4C-3-05 AUGMENTATION OF SUPPLY

(a) Any importation of water from outside the Basin shall be excluded from any allocation set forth elsewhere in this agreement or the Comprehensive Water

Management Plan, and the Party importing such water shall have the right to full and complete use and consumption of such imported water.

(b) Any Party implementing a conservation program with respect to water supplies shall be entitled to full and complete use and consumption of all increased supplies resulting from such conservation program. The burden of showing such increase shall rest on the Party claiming such increase.

Commentary: (1) Section 4C-3-06(a) makes clear that if a Party arranges to increase supplies by importing water, it need not share those additional supplies.

(2) Section 4C-3-06(b) provides encouragement for conservation by rewarding the Party which undertakes that effort. Caution should be exercised in incorporating this provision, however, inasmuch as the level of conservation efforts between the Parties may be unequal at the time the agreement is negotiated. A Party that has already made significant efforts should not be placed at a disadvantage relative to a Party that, prior to the agreement, made little effort to conserve.

Cross References: §2C-1-05 (powers of sovereign parties; withdrawal); §2C-2-02 (basin); §2C-2-03 (comprehensive water management plan); §2C-2-05 (drought); §3C-3-03 (projects of the signatory parties); §3C-3-04 (cooperative services); §4C-1-01 (joint exercise of sovereignty); §4C-1-02 (interrelationship of water resources); §4C-1-03 (comprehensive water management plan); §4C-1-04 (purpose and objectives of comprehensive water management plan); §4C-1-05 (conditions of comprehensive water management Plan); §4C-1-06 (deviation from comprehensive water management plan); §4C-1-08 (Allocation during drought conditions); §4C-1-09 (non-impairment of comprehensive water management plan by state action); §4C-2-01 (water allocation, generally); §4C-2-02 (waters not subject to allocation); §4C-2-03 (basin water sources); §4C-2-04 (basin water demands and needs);

Similar Agreements: None.

§4C-3-06 WATER QUALITY

<u>Alternative 1.</u>

The Commission shall:

(a) Manage the waters of the Basin to maintain ecosystem integrity, preserve and protect aquatic ecosystems effectively from any form of (significant) degradation on a drainage Basin or sub-Basin basis;

(b) Publish biological, health, physical and chemical quality criteria for all water bodies (surface and underground water), according to Basin capacities and needs, with a view to an ongoing improvement of water quality;

(c) Establish standards for the discharge of effluents and for the receiving waters, to include standards for land use management that relate to best management practices for water quality;

(d) Establish minimum flow criteria to maintain the desired instream environmental conditions and insure nourishment of wetlands and riparian buffers as necessary to properly filter nitrates and phosphorous arising from nonpoint runoff.

(e) Maintain the quality of the Waters of the Basin at or above water quality standards as may be adopted, now or hereafter, by the water pollution control agencies of the respective Parties in compliance with the provisions of the Federal Water Quality Act of 1965, and amendments thereto.

Alternative 2

(a) The Parties mutually agree to the principle of individual Party efforts to control natural and man-made water pollution within each Party and to the continuing support of both Parties in active water pollution control programs.

(b) The Parties agree to cooperate, through their appropriate Party agencies, in the investigation, abatement, and control of sources of alleged interparty pollution within the Basin.

(c) (Alternative 1). The Parties agree to cooperate in maintaining the quality of the Waters of the Basin at or above water quality standards as may be developed and agreed to by the Parties.

(c) (Alternative 2). The Parties agree to cooperate in maintaining the quality of the Waters of the Basin at or above water quality standards as may be adopted, now or hereafter, by the water pollution control agencies of the respective Parties in compliance with the provisions of the Clean Water Act of 1965, and amendments thereto.

Commentary. The quality of the water allocated is as important as the quantity of water allocated. Poor quality water imposes risks that the Parties should consider. First, there is the health risk to the population that uses the water for domestic purposes. Second, if the available water will not meet the standards for certain industrial purposes, there is the risk that economic growth will be impaired. Finally, there is the risk that quality degradation will have a severe impact on the ecology of the Basin, resulting in long-term sustainability complications. Integration of water quality and quantity is essential. Agenda 21 obligated all signatories to develop a program of water and sustainable development; *see also* A. Satre Ahlander, *Environmental Policies in the former Soviet Union*, Economic Policies for Sustainable Development, (Thomas Sterner, ed., 1994). The UN *Convention on the Law of the Non-Navigational Uses of International Watercourses*, United Nations Document A/51/869 (1998), establishes the criterion that "(w)atercourse States shall, individually and, where appropriate, jointly, protect and preserve the ecosystems of international watercourses."

Surface and underground water may be degraded by a variety of factors. Major problems affecting the quality of these water resources arise, for instance, from inadequate domestic sewage treatment, inadequate controls on the discharge of industrial waste and effluent, the diversion of waters resulting in insufficient water to assimilate waste, the loss and destruction of catchment areas, the improper siting of industrial plants, deforestation and poor agricultural practices which cause leaching of nutrients and pesticides. Transboundary water sharing must include effective plans and programs that eliminate, or at least minimize, the possible sources of water quality degradation. In most cases, water quality can be significantly advanced by the use of Best Management Practices. Such practices are forms of landuse management that limit non-point source pollution from entering water bodies. Examples include vegetative riparian buffers and limitations on contiguous impervious surfaces, among others. Best Management Practices are normally voluntary in nature but complemented with tax incentives. Alternative 1(c) provides the Commission with the power to establish such standards, although the Parties may choose to make the standards voluntary.

The complex interconnected nature of freshwater systems suggests that freshwater management should be systemically integrated, taking a catchment management approach which balances the needs of people and the environment. The Parties should manage the waters of the Basin to maintain ecosystem integrity, preserve aquatic ecosystems, and protect them effectively from any form of degradation on a drainage Basin or sub-Basin basis.

The Parties should establish biological, health, physical and chemical quality criteria for all significant water bodies in the Basin to continually improve water quality. The Parties should establish minimum standards both for discharging effluents and for receiving waters. We recommend the Parties institute standards for land use management such as limits on agrochemical use, deforestation, and wasteful irrigation practices. Such rational land use standards should prevent land degradation, erosion and siltation of lakes and other water bodies.

Cross-references: §1C-1-01 (general policies); §1C-1-02 (purposes of agreement); §1C-1-03 (objectives of agreement); §2C-2-02 (basin); §2C-2-03 (comprehensive water management plan); §2C-2-05 (drought); §2C-2-09 (party or parties); §2C-2-10 (underground water); §2C-2-11 (waters of the basin); §3C-3-03 (projects of the signatory parties); §4C-1-01 (joint exercise of sovereignty); §4C-1-02 (interrelationship of water resources); §4C-1-03 (comprehensive water management plan); §4C-1-04 (purpose and objectives of comprehensive water management plan); §4C-1-05 (conditions of comprehensive water management Plan); §4C-1-06 (deviation from comprehensive water management plan); §4C-1-09 (non-impairment of comprehensive water management plan by state action); §4C-2-01 (water allocation, generally); §4C-2-02 (waters not subject to allocation); §4C-2-03 (basin water sources); §4C-2-04 (basin water demands and needs); §4C-2-05 (allocation to equitable and reasonable uses); §4C-2-06 (watershed management).

Similar Agreements: *Rio Grande Compact*, 53 Stat. 785 (1939); *Klamath River Basin Compact*, 71 Stat. 497 (1957); *Arkansas River Basin Compact of 1965*, 80 Stat. 1409 (1966); *Big Blue River Basin Compact*, 86 Stat. 193 (1972); *Arkansas River Basin Compact of 1970*, 87 Stat. 569 (1973); *Delaware River Basin Compact*, Pub. L. 87-328, 75 Stat. 688 (1961); *Susquehanna River Basin Compact*, Pub. L. No. 91-575, 84 Stat. 1509 (1970); *Kansas-Nebraska Big Blue*

River Compact, 86 Stat. 193 (1972); *Red River Compact*, 94 Stat. 3305 (1980); Oregon-California Goose Lake Interstate Compact, 98 Stat. 291 (1984); *Treaty between the United States and Great Britain relating to Boundary Waters, and Questions arising between the United States and Canada*, 36 Stat. 2451 (1909); *Agreement between the United States and Canada on Great Lakes Water Quality*, 1153 UNTS 187 (1978); *Convention on the Protection and Use of Transboundary Watercourses and International Lakes*, 31 I.L.M. 1312 (1992); *Convention on the Law of the Non-Navigational Uses of International Watercourses*, United Nations Document A/51/869 (1998).

§4C-3-07 UNDERGROUND WATER; LIMIT ON WITHDRAWALS (Optional)

Alternative 1

When such action is necessary [to maintain an allocation set out elsewhere], [Upstream Party] shall regulate, in the same manner that surface flow is regulated, withdrawal of water from irrigation wells located within ___ miles of the river or its tributaries.

Alternative 2

When such action is necessary [to maintain an allocation set out elsewhere], [Upstream Party] shall regulate, in the same manner that surface flow is regulated, withdrawal of water from irrigation wells that may be hydrologically connected to the river or its tributaries.

Commentary: If underground water is subject to the allocation provisions of the agreement, it may be useful to specifically address the steps to be taken with respect to withdrawals. Alternative 1, adapted from the *Big Blue River Compact*, 86 Stat. 193 (1972) uses a distance limitation to determine which wells fall within the scope of the agreement. If it is possible to establish the hydrological connection between all wells and the surface flow, the mileage limitation may be replaced with references to wells with such connection. In the absence of definitive hydrologic information, the mileage limitation may make administration easier, if less precise.

No further specific allocation systems for underground water are provided because it is assumed that if underground water is allocated by agreement, that allocation will be in conjunction with allocation of related surface water sources and the allocation of underground water will be incorporated as part of the overall allocation of water. If underground water is allocated independently from surface water, the parties might use the surface models as a guide with respect to types of allocations (proportional, guaranteed minimum, etc. - *See* Model B)

Cross References: §1C-1-01 (general policies); §1C-1-02 (purposes of agreement); §1C-1-03 (objectives of agreement); §2C-2-03 (comprehensive water management plan); §2C-2-05 (drought); §2C-2-09 (party or parties); §2C-2-10 (underground water); §4C-1-01 (joint exercise of sovereignty); §4C-1-02 (interrelationship of water resources); §4C-1-03 (comprehensive water management plan); §4C-1-04 (purpose and objectives of comprehensive water management plan); §4C-1-05 (conditions of comprehensive water management Plan); §4C-1-06 (deviation from comprehensive water management plan); §4C-1-01 (joint exercise of sovereignty); §4C-1-02 (interrelationship of water resources); §4C-1-03 (comprehensive water management plan); §4C-1-04 (purpose and objectives of comprehensive water management plan); §4C-1-05 (conditions of comprehensive water management Plan); §4C-1-06 (deviation from comprehensive water management plan); §4C-1-07 (allocation during flood conditions); §4C-1-08 (Allocation during drought conditions); §4C-1-09 (non-impairment of comprehensive water management plan by state action); §4C-2-01 (water allocation, generally); §4C-2-02 (waters not subject to allocation); §4C-2-03 (basin water sources); §4C-2-04 (basin water demands and needs); §4C-2-05 (allocation to equitable and reasonable uses); §4C-2-06 (watershed management); 4-3-02, Withdrawals and Diversion; Protected Areas; §4C-3-05 (water levels protected); §4C-3-06 (augmentation of supply).

Similar Agreements: *Kansas-Nebraska Big Blue River Compact*, 86 Stat. 193 (1972).

§4C-3-08   ATMOSPHERIC WATER

Any Party which augments precipitation within the Basin shall be entitled to full and exclusive use of additional water supplies resulting from such augmentation, notwithstanding any other standard of allocation set forth in this agreement or the Comprehensive Water Management Plan. In the event the Parties cannot agree on whether or to what extent precipitation has been augmented, the Party asserting the right to such increased supplies shall bear the burden of proving that the increase, if any, was the result of the Party's augmentation efforts and not simply the result of natural variation in precipitation amounts.

Commentary: As technology advances, precipitation augmentation may become more widely practiced an accepted. This provision would allow a Party that financed such efforts to retain the benefits, provided it could meet the burden of proof. Absent inclusion of this alternative, the augmented supply will be treated in the same manner as any other water within the Basin.

Cross References: §1C-1-01 (general policies); §1C-1-02 (purposes of agreement); §1C-1-03 (objectives of agreement); §2C-2-01 (atmospheric water); §2C-2-03 (comprehensive water management plan); §2C-2-05 (drought); §2C-2-09 (party or parties); §4C-1-01 (joint exercise of sovereignty); §4C-1-02 (interrelationship of water resources); §4C-1-03 (comprehensive water management plan); §4C-1-04 (purpose and objectives of comprehensive water management plan); §4C-1-05 (conditions of comprehensive water management Plan); §4C-1-06 (deviation from comprehensive water management plan); §4C-1-01 (joint exercise of sovereignty); §4C-1-02 (interrelationship of water resources); §4C-1-03 (comprehensive water management plan); §4C-1-04 (purpose and objectives of comprehensive water management plan); §4C-1-05 (conditions of comprehensive water management Plan); §4C-1-06 (deviation from comprehensive water management plan); §4C-1-07 (allocation during flood conditions); §4C-1-08 (Allocation during drought conditions); §4C-1-09 (non-impairment of comprehensive water management plan by state action); §4C-2-01 (water allocation, generally); §4C-2-02 (waters not subject to allocation); §4C-2-03 (basin water sources); §4C-2-04 (basin water demands and needs); §4C-2-05 (allocation to equitable and reasonable uses); §4C-2-06 (watershed management); 4-3-02, Withdrawals and Diversion; Protected Areas; §4C-3-05 (water levels protected); §4C-3-06 (augmentation of supply).

Similar Agreements: *Rio Grande Compact*, 53 Stat. 785 (1938).

§4C-3-09   RECREATION (Optional)

(a) The Commission shall provide for the development of water related public sports and recreational facilities. The Commission on its own account or in cooperation with a signatory Party, political subdivision or any agency thereof, may provide for the construction, maintenance and administration of such facilities.

(b) The Commission shall not operate any such project or facility unless it has first found and determined that no other suitable unit or agency of government is available to operate the same upon reasonable conditions, in accordance with the intent and purpose expressed in §2C-1-06 of this Agreement.

(c) The Commission shall encourage activities of other public agencies having water related recreational interests and assist in the coordination thereof.

(d) The Commission shall recommend standards for the development and administration of water related recreational facilities.

(e) The Commission shall provide for the administration, operation and maintenance of recreational facilities owned or controlled by the Commission and for the letting and supervision of private concessions in accordance with this article.

(f) The Commission shall, after notice and public hearing, provide by regulation for the award of contracts for private concessions in connection with recreational facilities, including any renewal or extension thereof, upon sealed competitive bids after public advertisement therefore.

Commentary: This section may be appropriate for agreements involving developed nations. However, the issue may not be material to agreements involving developing nations, or those in transition.

Cross-references: §1C-1-01 (general policies); §1C-1-02 (purposes of agreement); §1C-1-03 (objectives of agreement); §2C-2-03 (comprehensive water management plan); §2C-2-09 (party or parties); §4C-1-01 (joint exercise of sovereignty); §4C-1-02 (interrelationship of water resources); §4C-1-03 (comprehensive water management plan); §4C-1-04   (purpose and objectives of comprehensive water management plan); §4C-1-05 (conditions of comprehensive water management Plan); §4C-1-06 (deviation from comprehensive water management plan); §4C-1-07 (allocation during flood conditions); §4C-1-08 (Allocation during drought conditions); §4C-1-09 (non-impairment of comprehensive water management plan by state action); §4C-2-01 (water allocation, generally); §4C-2-02 (waters not subject to allocation); §4C-2-03 (basin water sources); §4C-2-04 (basin water demands and needs); §4C-2-05 (allocation to equitable and reasonable uses); §4C-2-06 (watershed management); §4C-3-01 (existing rights recognized); §4C-3-02 (flood protection works); §4C-3-03 (minimum flows); §4C-3-04 (withdrawals and diversion; protected areass); §4C-3-05 (water levels protected).

Similar Agreements: *Delaware River Basin Compact*, Pub. L. 87-328, 75 Stat. 688 (1961); *Susquehanna River Basin Compact*, Pub. L. No. 91-575, 84 Stat. 1509 (1970).

ARTICLE 5C

FINANCING

<u>Alternative 1</u> **(Simplified)**

The budget of the Commission shall be prepared by the Executive Director and approved by unanimous vote of the Commissioners.

<u>Alternative 2</u> **(Comprehensive)**

The Commission shall annually adopt a current expense budget and a capital budget that includes any and all capital projects it proposes to undertake or continue during the budget period. The amount required to balance the current expense budget and the capital budget will be financed according to the arrangements described in Annex C-1.

Commentary: Comprehensive planning and management requires dependable sources of finances. If a program of financing is uncertain, or if the program is not sufficiently detailed to provide certainty, comprehensive planning and management will fail. However, in some situations, the Parties may wish to minimize complexities and maximize flexibility on the part of the Commission. Therefore, two alternative financing methodologies are provided for consideration.

Alternative 1 is a simple methodology that leaves funding of the agreement implementation to predetermined shares from the Parties. The provision provided above suggests equality of contribution. Equally valid would be contributions on an equitable basis, where "equitable" is determined according to the economic benefit received as the result of the comprehensive water management. The contributions could also be based on the percentage of basin within each of the Parties. Other possibilities include the relative wealth of the Parties as expressed in GNP or the relative wealth of various reaches of the watercourses within the Parties. It should be recognized that such simplified funding exposes the comprehensive management to the vagaries of political forces within the various sovereign Parties.

Alternative 2 provides a methodology that is more comprehensive in nature and which provides a great degree of independence from political forces that may arise in one or more of the Parties. It is, however, detailed and appropriate only where the Parties to the Agreement do not impose significant control over economic policies and individual and corporate decision-making. This alternative is drawn, largely unmodified, from the *Delaware River Basin Compact* and its amendments. They are not appropriate in all situations, being oriented towards a federal political system.

Cross-references: §1C-1-01 (general policies); §1C-1-02 (purposes of agreement); §1C-1-03 (objectives of agreement); §1C-1-04 (coordination and cooperation); §1C-1-05 (reservation of federal rights); §1C-1-06 (national security); §2C-1-01 (activation of agreement); §2C-1-02 (consent to jurisdiction); §2C-1-03 (duration of agreement); §2C-1-04 (amendments and supplements); §2C-1-05 (powers of sovereign parties; withdrawal); §2C-1-06 (existing agencies); §2C-1-07 (limited applicability); §2C-1-08 (annexes); §2C-2-02 (basin); §2C-2-03, Conservation Measures; §2C-2-04, Consumptive Use; §2C-2-05, Domestic Use; §2C-2-06, Drought Management Strategies; §2C-2-07, Equitable Approtionment; §2C-2-08, Equitable and Reasonable Utilization; §2C-2-09, Interbasin Transfer; §2C-2-10, Municipal Uses; §2C-2-11, Nonconsumptive Uses; §2C-2-12, Party; §2C-2-13, Underground Water; §2C-2-14, Waters of the Basin; §3C-1-01 (commission created); §3C-1-02 (jurisdiction of the commission); §3C-1-03 (commissioners); §3C-1-04 (status, immunities and privileges); §3C-1-05 (commission organization and staffing); §3C-1-06 (rules of procedures); §3C-1-07 (commission administration); §3C-2-01 (general powers and duties); §3C-2-02 (powers and duties reserved to the commissioners); §3C-2-03 (obligations of the commission); §3C-2-04 (regulations; enforcement); §3C-2-05 (prohibited activities); §3C-2-06 (referral and review); §3C-2-07 (advisory committees); §3C-2-08 (reports); §3C-2-09 (condemnation proceedings); §3C-2-10 (meetings, hearings and records); §3C-2-11 (tort liability); §3C-3-01 (coordination and cooperation); §3C-3-02 (project costs and evaluation standards); §3C-3-03 (projects of the signatory parties); §3C-3-04 (cooperative services).

Similar Agreements: *Delaware River Basin Compact*, Pub. L. 87-328, 75 Stat. 688 (1961); *Klamath River Basin Compact*, 71 Stat. 497 (1957); The Mekong River Basin Treaty, 34 ILM 864 (1995); *Susquehanna River Basin Compact*, Pub. L. No. 91-575, 84 Stat. 1509 (1970).

ARTICLE 6C

DISPUTE RESOLUTION

Commentary. Disputes will inevitably arise as the Agreement is implemented. Thesed disputes may involve differences in interpretation of the Agreement's provisions or non-compliance with the Agreement itself. The disputes may also arise because of changing conditions that alter the effectiveness of the Agreement for one or more of the Parties. While a speedy and equitable process of dispute resolution serves all Parties well, some sovereign entities do not wish to enter into an obligatory process. In such a case, Article 6C may be omitted. In other instances, the Parties may recognize the need to institutionalize a dispute resolution process

§6C-1-01 GOOD FAITH IMPLEMENTATION

Each of the Parties pledges to support implementation of all provisions of this Agreement, and covenants that its officers and agencies will not hinder, impair, or prevent any other Party carrying out any provision of this Agreement.

Commentary. In the negotiations, each party seeks the rights and authorities critical to certain political, economic or social objectives while ceding less critical rights and authorities to the other nations. While accepting this fact, under the *Convention on the Law of the Non-Navigational Uses of International Watercourses*, United Nations Document A/51/869 (1998), all Parties have a duty to cooperate and negotiate in good faith. This principle is the foundation of international law, and it applies in all relations between sovereign states. All states are expected to conduct themselves with an absence of malice and with no intention to seek unconscionable advantage, or otherwise be deceitful.

It must be noted, however, that good-faith misinterpretation of compact obligations do not excuse a Party from damage liability. *See Texas v. New Mexico*, 482 U.S. 124 (1987). In that case, the Supreme Court reasoned that a compact is a contract, and standard contract law does not allow a defense based on misinterpretation of contract obligations. *See* Grant, §45.07(c), §46.05(d).

Cross-references: §1C-1-01 (general policies); §1C-1-02 (purposes of agreement); §1C-1-03 (objectives of agreement); §1C-1-04 (coordination and cooperation); §1C-1-05 (reservation of federal rights); §1C-1-06 (national security); §2C-1-01 (activation of agreement); §2C-1-02 (consent to jurisdiction); §2C-1-03 (duration of agreement); §2C-1-04 (amendments and supplements); §2C-1-05 (powers of sovereign parties; withdrawal); §2C-1-06 (existing agencies); §2C-1-07 (limited applicability); §2C-1-08 (annexes); §2C-2-09 (party or parties); §3C-1-01 (commission created); §3C-1-02 (jurisdiction of the commission); §3C-1-03 (commissioners); §3C-1-04 (status, immunities and privileges); §3C-1-05 (commission organization and staffing); §3C-1-06 (rules of procedures); §3C-1-07 (commission administration); §3C-2-01 (general powers and duties); §3C-2-02 (powers and duties reserved to the commissioners); §3C-2-03 (obligations of the commission); §3C-2-04 (regulations; enforcement); §3C-2-05 (prohibited activities); §3C-2-06 (referral and review); §3C-2-07 (advisory committees); §3C-2-08 (reports); §3C-2-09 (condemnation proceedings); §3C-2-10 (meetings, hearings and records); §3C-2-11 (tort liability); §3C-3-01 (coordination and cooperation); §3C-3-02 (project costs and evaluation standards); §3C-3-03 (projects of the signatory parties); §3C-3-04 (cooperative services); §4C-1-01 (joint exercise of sovereignty); §4C-1-02 (interrelationship of water resources); §4C-1-03 (comprehensive water management plan); §4C-1-04 (purpose and objectives of comprehensive water management plan); §4C-1-05 (conditions of comprehensive water management Plan); §4C-1-06 (deviation from comprehensive water management plan); §4C-1-07 (allocation during flood conditions); §4C-1-08 (Allocation during drought conditions); §4C-1-09 (non-impairment of comprehensive water management plan by state action); §4C-2-01 (water allocation, generally); §4C-2-02 (waters not subject to allocation); §4C-2-03 (basin water sources); §4C-2-04 (basin water demands and needs); §4C-2-05 (allocation to equitable and reasonable uses); §4C-2-06 (watershed management); §4C-3-01 (existing rights recognized); §4C-3-02 (flood protection works); §4C-3-03 (minimum flows); §4C-3-04 (withdrawals and diversion; protected areass); §4C-3-05 (water levels protected); §4C-3-06 (augmentation of supply); §4C-3-07 (water quality); §4C-3-08 (underground water; limit on withdrawals); §4C-3-09 (atmospheric water); §4C-3-10 (recreation); Article 5C (financing); §6C-1-01 (good faith implementation).

Similar Agreements: *Convention on the Law of the Non-Navigational Uses of International Watercourses*, United Nations Document A/51/869 (1998); *Charter of Economic Rights and Duties of States*, 1975); *Helsinki Rules on the Uses of the Waters of International Rivers*, 52 I.L.A. 484 (1966); *Stockholm Declaration of the United Nations Conference on the Human Environment*, 11 I.L.M. 1416 (1972).

§6C-1-02 MODIFICATION OF AGREEMENT

In the event that any Party is substantially hindered or prevented from performing any obligation or implementing any provision under this Agreement, by reasons of circumstances beyond the control of the Party (including but not limted to Acts of God, natural disasters, or labor disputes), the Parties agree to meet and negotiate an appropriate modification of the applicable provisions of the Agreement to reflect the effect of such force majeure. Such modifications may include extensions of applicable schedules and timetables, or agreements on substitute actions to fulfill the objectives and spirit of this Agreement.

Commentary: This provision provides a remedy for unintentional breaches of the Agreement that may occur due to unforeseen situations or changed conditions.

Cross-references: §1C-1-01 (general policies); §1C-1-02 (purposes of agreement); §1C-1-03 (objectives of agreement); §1C-1-04 (coordination and cooperation); §1C-1-05 (reservation of federal rights); §1C-1-06 (national security); §2C-1-01 (activation of agreement); §2C-1-02 (consent to jurisdiction); §2C-1-03 (duration of agreement); §2C-1-04 (amendments and supplements); §2C-1-05 (powers of sovereign parties; withdrawal); §2C-1-06 (existing agencies); §2C-1-07 (limited applicability); §2C-1-08 (annexes); §2C-2-09 (party or parties); §3C-1-01 (commission created); §3C-1-02 (jurisdiction of the commission); §3C-1-03 (commissioners); §3C-1-04 (status, immunities and privileges); §3C-1-05 (commission organization and staffing); §3C-1-06 (rules of procedures); §3C-1-07 (commission administration); §3C-2-01 (general powers and duties); §3C-2-02 (powers and duties reserved to the commissioners); §3C-2-03 (obligations of the commission); §3C-2-04 (regulations; enforcement); §3C-2-05 (prohibited activities); §3C-2-06 (referral and review); §3C-2-07 (advisory committees); §3C-2-08 (reports); §3C-2-09 (condemnation proceedings); §3C-2-10 (meetings, hearings and records); §3C-2-11 (tort liability); §3C-3-01 (coordination and cooperation); §3C-3-02 (project costs and evaluation standards); §3C-3-03 (projects of the signatory parties); §3C-3-04 (cooperative services); §4C-1-01 (joint exercise of sovereignty); §4C-1-02 (interrelationship of water resources); §4C-1-03 (comprehensive water management plan); §4C-1-04 (purpose and objectives of comprehensive water management plan); §4C-1-05 (conditions of comprehensive water management Plan); §4C-1-06 (deviation from comprehensive water management plan); §4C-1-07 (allocation during flood conditions); §4C-1-08 (Allocation during drought conditions); §4C-1-09 (non-impairment of comprehensive water management plan by state action); §4C-2-01 (water allocation, generally); §4C-2-02 (waters not subject to allocation); §4C-2-03 (basin water sources); §4C-2-04 (basin water demands and needs); §4C-2-05 (allocation to equitable and reasonable uses); §4C-2-06 (watershed management); §4C-3-01 (existing rights recognized); §4C-3-02 (flood protection works); §4C-3-03 (minimum flows); §4C-3-04 (withdrawals and diversion; protected areass); §4C-3-05 (water levels protected); §4C-3-06 (augmentation of supply); §4C-3-07 (water quality); §4C-3-08 (underground water; limit on withdrawals); §4C-3-09 (atmospheric water); §4C-3-10 (recreation); Article 5C (financing); §6C-1-01 (good faith implementation).

Similar Agreements: *Agreement on the Full Utilization of the Nile Waters, Nov. 8, 1959, UAR-Sudan,* UNTS Vol. 453, 51.; *Vienna Convention for the Protection of the Ozone Layer,* 26 I.L.M. 1529 (1985); *Delaware River Basin Compact,* Pub. L. 87-328, 75 Stat. 688 (1961); *Susquehanna River Basin Compact,* Pub. L. No. 91-575, 84 Stat. 1509 (1970).

§6C-1-03   MATERIAL BREACH

The Parties consider this agreement to be comlete and an integral whole. Each recommendation and provision of this agreement is considered material to the entire Agreement, and failure to implement or adhere to any recommendation or provision may be considered a material breach.

Commentary. This provision is standard to most contractual agreements, indicating that all provisions of the Agreement are interrelated and that breach of any provision by one Party may void the Agreement.

Cross-references: §1C-1-01 (general policies); §1C-1-02 (purposes of agreement); §1C-1-03 (objectives of agreement); §1C-1-04 (coordination and cooperation); §1C-1-05 (reservation of federal rights); §1C-1-06 (national security); §2C-1-01 (activation of agreement); §2C-1-02 (consent to jurisdiction); §2C-1-03 (duration of agreement); §2C-1-04 (amendments and supplements); §2C-1-05 (powers of sovereign parties; withdrawal); §2C-1-06 (existing agencies); §2C-1-07 (limited applicability); §2C-1-08 (annexes); §2C-2-09 (party or parties); §3C-1-01 (commission created); §3C-1-02 (jurisdiction of the commission); §3C-1-03 (commissioners); §3C-1-04 (status, immunities and privileges); §3C-1-05 (commission organization and staffing); §3C-1-06 (rules of procedures); §3C-1-07 (commission administration); §3C-2-01 (general powers and duties); §3C-2-02 (powers and duties reserved to the commissioners); §3C-2-03 (obligations of the commission); §3C-2-04 (regulations; enforcement); §3C-2-05 (prohibited activities); §3C-2-06 (referral and review); §3C-2-07 (advisory committees); §3C-2-08 (reports); §3C-2-09 (condemnation proceedings); §3C-2-10 (meetings, hearings and records); §3C-2-11 (tort liability); §3C-3-01 (coordination and cooperation); §3C-3-02 (project costs and evaluation standards); §3C-3-03 (projects of the signatory parties); §3C-3-04 (cooperative services); §4C-1-01 (joint exercise of sovereignty); §4C-1-02 (interrelationship of water resources); §4C-1-03 (comprehensive water management plan); §4C-1-04 (purpose and objectives of comprehensive water management plan); §4C-1-05 (conditions of comprehensive water management Plan); §4C-1-06 (deviation from comprehensive water management plan); §4C-1-07 (allocation during flood conditions); §4C-1-08 (Allocation during drought conditions); §4C-1-09 (non-impairment of comprehensive water management plan by state action); §4C-2-01 (water allocation, generally); §4C-2-02 (waters not subject to allocation); §4C-2-03 (basin water sources); §4C-2-04 (basin water demands and needs); §4C-2-05 (allocation to equitable and reasonable uses); §4C-2-06 (watershed management); §4C-3-01 (existing rights recognized); §4C-3-02 (flood protection works); §4C-3-03 (minimum flows); §4C-3-04 (withdrawals and diversion; protected areass); §4C-3-05 (water levels protected); §4C-3-06 (augmentation of supply); §4C-3-07 (water quality); §4C-3-08 (underground water; limit on withdrawals); §4C-3-09 (atmospheric water); §4C-3-10 (recreation); Article 5C (financing); §6C-1-01 (good faith implementation).

Similar Agreements: *Delaware River Basin Compact*, Pub. L. 87-328, 75 Stat. 688 (1961); *Susquehanna River Basin Compact*, Pub. L. No. 91-575, 84 Stat. 1509 (1970).

§6C-1-04 NEGOTIATIONS AND CONSULTATIONS

(a) If any Party believes another Party has violated or failed to carry out any provision of this Agreement, it shall notify such Party, and all other Parties, in writing specifying the alleged violation or failure.

(b) The complaining Party shall notify the Commission of the dispute and the intention to enter into negotiations and consultations.

(c) Within (___) days of notice provided under paragraph (a), all Parties will meet to discuss the alleged violation or failure and to negotiate an appropriate settlement, including actions to correct such violation or failure. Such discussions and negotiations shall be pursued in good faith for not less than (___) days after original notice.

(d) The Parties shall seek to avoid any resolution that adversely affects the interests under this Agreement of any other Party.

Commentary. Negotiation is a process in which the conflicting parties engage in face-to-face discussions to develop a mutually satisfactory agreement on the issues or problems at hand. No outside, independent party or individual is involved. If negotiations between the parties themselves are not effective, the process evolves into mediation.

Cross-references: §1C-1-01 (general policies); §1C-1-02 (purposes of agreement); §1C-1-03 (objectives of agreement); §1C-1-04 (coordination and cooperation); §2C-1-02 (consent to jurisdiction); §2C-1-03 (duration of agreement); §2C-1-04 (amendments and supplements); §2C-1-05 (powers of sovereign parties; withdrawal); §2C-2-09 (party or parties); §4C-1-01 (joint exercise of sovereignty).

Similar Agreements: *Convention on the Law of the Sea*, 33 I.L.M. 1309 (1982); *Convention on the Law of the Non-Navigational Uses of International Watercourses*, United Nations Document A/51/869 (1998); *Convention on the Protection and Use of Transboundary Watercourses and International Lakes*, 31 I.L.M. 1312 (1992); *North American Free Trade Agreement* (1993); *Treaty of Peace between the State of Israel and the Hashemite Kingdom of Jordan*, 34 I.L.M. 43 (1994); *Apalachicola-Chattahoochee-Flint River Basin Compact*, O.C.G.A. 12-10-100 (1997); *Alabama-Coosa-Tallapoosa River Basin Compact*, O.C.G.A. 12-10-110 (1997); *Delaware River Basin Compact*, Pub. L. 87-328, 75 Stat. 688 (1961); *Susquehanna River Basin Compact*, Pub. L. No. 91-575, 84 Stat. 1509 (1970).

Alternative 1 (Right to Litigate)

§6-1-05   RIGHT TO LITIGATE   (Optional)

Nothing in this agreement shall be construed to limit or prevent either Party from instituting or maintaining any action or proceeding, legal or equitable, in any tribunal of competent jurisdiction for the protection of any right under this agreement or the enforcement of any of its provisions.

Commentary: If the Parties choose not to adopt a provision that states that the decision of the Arbitral Panel shall be considered final and shall not be appealable to any court of law or equity [*see* § 6C-1-05, *et seq*], the Agreement must provide for appropriate legal or equitable action or proceeding. The existence of an appropriate tribunal may pose a problem in cases not involving an entity like the United States or European Union. It may be advisable to specify the tribunal in the agreement itself to avoid dispute over jurisdictional questions at a later date.

Cross References: §1C-1-01 (general policies); §1C-1-02 (purposes of agreement); §1C-1-03 (objectives of agreement); §1C-1-04 (coordination and cooperation); §2C-1-02 (consent to jurisdiction); §2C-1-03 (duration of agreement); §2C-1-04 (amendments and supplements); §2C-1-05 (powers of sovereign parties; withdrawal); §2C-2-09 (party or parties); §4C-1-01 (joint exercise of sovereignty).

Similar Agreements: *Belle Fourche River Compact*, 58 Stat. 94 (1944); *Colorado River Compact*, 45 Stat. 1057 (1928); *Snake River Compact*, 64 Stat. 29 (1949).

Alternative 2 (Alternative Dispute Resolution)

§6C-1-05   ALTERNATIVE DISPUTE RESOLUTION

(a) Desiring that this Agreement be carried out in full, the Parties agree that disputes between the Parties regarding interpretation, application; and implementation of this Agreement shall be settled by alternative dispute resolution and agree to forswear litigation.

(b) The dispute settlement provisions of this Article shall apply with respect to the avoidance or settlement of all disputes between the Parties regarding the interpretation or application of this Agreement or wherever a Party considers that an actual or proposed measure of another Party is or would be inconsistent with the obligations of this Agreement or cause nullification or impairment of the application of the Agreement.

Commentary. Disputes will inevitably arise as the Agreement is implemented and enforced. They may involve differences in interpretation of the Agreement's provisions or non-compliance with the Agreement itself. The disputes may also arise because of changing conditions that alter the effectiveness of the Agreement for one or more of the Parties. Therefore, the institutional provisions should provide for a process to resolve disputes quickly, effectively and permanently. The mechanism should emphasize a streamlined process of dispute resolution that minimizes costly, time-consuming litigation. Alternate dispute resolution (ADR) is a major force in the resolution of disputes about the terms of an international agreement. This is necessarily so because of the lack of effective supranational mechanisms for jurisprudence. Even where judicial remedies are available, however, ADR is may be a preferable way to resolve disputes.

Judicial conflict resolution holds significant disadvantages. (a) Judges are generalists who are, in most cases, dependent on the testimony presented before them. (b) The judicial process, with its manifold procedural safeguards, is too slow for effective natural resources management. (c) Judicial decrees are retrospective, geographically limited and quite fact-specific. Unlike administrative agencies, courts are incapable of issuing prospective, uniform regulations of general applicability. (d) Courts lack the ability to consistently monitor and evaluate solutions they have devised. (e) Water is a public trust resource that should be managed by institutions that are politically responsive to the public. (William Goldfarb, *The Trend Toward Judicial Integration of Water Quality and Quantity Management: Facing the New Century* in Water Resources Administration in the United States (M. Reuss, ed., 1993).

Alternative dispute resolution (ADR) processes are designed to resolve disputes as quickly as possible and at the lowest cost to the Parties involved. The process consists of a successive series of techniques that become increasingly time-consuming and expensive. These techniques are negotiation, mediation, arbitration and litigation. With each successive step, the Parties spend more time and more money for a result over which the Parties have less and less control.

The agreement to forswear litigation appearing in §6C-1-04 is optional. If this phrase is included and §6C-1-08 is not included, judicial intervention will not obtain. *See* Grant, § 46.05 discussing the Supreme Court's statements in *Texas v. New Mexico*, 462 U.S. 540 (1980).

Cross-references: 1-1-01, General Policies; §1C-1-02 (purposes of agreement); §1C-1-03 (objectives of agreement); §1C-1-04 (coordination and cooperation); §2C-1-02 (consent to jurisdiction); §2C-1-03 (duration of agreement); §2C-1-04 (amendments and supplements); §2C-1-05 (powers of sovereign parties; withdrawal); §2C-2-09 (party or parties); §4C-1-01 (joint exercise of sovereignty).

Similar Agreements: *Agreement on the Cooperation for the Sustainable Development of the Mekong River Basin*, 34 ILM 864 (1995); *Convention on the Law of the Non-Navigational Uses of International Watercourses*, United Nations Document A/51/869 (1998); *Convention on the*

Law of the Sea, 33 I.L.M. 1309 (1982); *Convention on the Protection and Use of Transboundary Watercourses and International Lakes*, 31 I.L.M. 1312 (1992); *North American Free Trade Agreement*, 19 U.S.C. §§ 3311-3473 (1993); *Delaware River Basin Compact*, Pub. L. 87-328, 75 Stat. 688 (1961); *Susquehanna River Basin Compact*, Pub. L. No. 91-575, 84 Stat. 1509 (1970).

§6C-1-06 CONCILIATION AND MEDIATION

(a) If the Parties are unable to reach agreement on a settlement, after good faith discussions and negotiations within the period provided in §6C-1-05(c) any aggrieved Party may request that the Commission institute measures of conciliation and mediation. The requesting Party shall state in the request the nature of the dispute and indicate the provisions of the Agreement that it considers relevant, and shall deliver the request to the other Parties, and to the Commission. Unless it decides otherwise, the Commission shall convene within (___) days of delivery of the request and shall endeavor to resolve the dispute promptly.

(b) The Commission may call on such technical advisers or create such working groups or expert groups as it deems necessary to act as mediators and assist in the resolution of the dispute and make recommendations as may assist the consulting Parties to reach a mutually satisfactory resolution of the dispute.

Commentary. Mediation is the intervention of a "third party" in a dispute between two other parties in an attempt to reconcile their differences, usually upon their request. The qualifications of the mediator require his or her knowledge, experience and background in the water resource issues themselves as well as an understanding and background in the legal issues involved. The mediation moves through three stages:

(a) The mediator identifies and develops a factual discussion of the disputed and undisputed issues with the parties, both individually and collectively. The purpose is to assure that all parties understand the strengths and weaknesses of their case and the perceived weaknesses and strengths of the opposing parties.

(b) The mediator explores with the parties their goals, objectives and interests, attempting to create alternative solutions to their perceived concerns. This portion of the process also involves discussions with the parties individually and collectively.

(c) After the mediator intervenes, the parties themselves may then reassume a negotiating posture and possibly agree on a mutually acceptable alternative solution, or they may proceed to arbitration.

Cross-references: §1C-1-01 (general policies); §1C-1-02 (purposes of agreement); §1C-1-03 (objectives of agreement); §1C-1-04 (coordination and cooperation); §2C-1-02 (consent to jurisdiction); §2C-1-03 (duration of agreement); §2C-1-04 (amendments and supplements); §2C-1-05 (powers of sovereign parties; withdrawal); §2C-2-09 (party or parties); §4C-1-01 (joint exercise of sovereignty).

Similar Agreements: *Agreement on Cooperation for the Sustainable Development of the Mekong River Basin*, 34 ILM 864 (1995); *Convention on the Law of the Non-Navigational Uses of International Watercourses*, United Nations Document A/51/869 (1998); *Convention on the Protection and Use of Transboundary Watercourses and International Lakes*, 31 I.L.M. 1312 (1992); *Treaty of Peace between the State of Israel and the Hashemite Kingdom of Jordan*, 34 I.L.M. 43 (1994); Apalachicola- -Flint River Basin Compact, O.C.G.A. 12-10-100 (1997); *Alabama-Coosa-Tallapoosa River Basin Compact*, O.C.G.A. 12-10-110 (1997); *Delaware River Basin Compact*, Pub. L. 87-328, 75 Stat. 688 (1961); *Susquehanna River Basin Compact*, Pub. L. No. 91-575, 84 Stat. 1509 (1970).

§6C-1-07 ARBITRATION

(a) If the Commission has instituted mediation efforts as described in §6C-1-06 and the matter has not been resolved within 60 days thereafter any consulting Party may request in writing the establishment of an arbitral panel. The requesting Party shall deliver the request to the other Parties and the Commission.

(b) On delivery of the request, the Commission shall establish an arbitral panel.

(c) Unless otherwise agreed by the disputing Parties, the panel shall be established and perform its functions in a manner consistent with the provisions of Annex C-2.

(d) (Optional) The decision of the Arbitral Panel shall be considered final and shall not be appealable to any court of law or equity.

Commentary. Essentially arbitration is an informal trial, with the parties choosing the judge in a process that has a less formal evidentiary process. Arbitration differs from mediation only in the rules of decision as to the solution chosen to resolve the dispute. Whereas the mediator seeks to persuade the parties to agree on a mutually acceptable solution, the parties agree to allow the arbitrator to make decisions that are binding on the parties.

It should be recognized that implementation and enforcement of the Agreement is as important as the actual development. Most dispute resolution will most likely be accomplished through Arbitral Panels. Unless the Parties ensure effective dispute resolution, the Agreement may very well become meaningless. If the Parties decide to eliminate this detail, it is recommended that a side agreement be formed that describes the dispute resolution mechanism. The main Agreement should make reference to the side Agreement. In any event, the standards presented for the Arbitral Panel should insure that the members of the panel are both technically qualified in the subject of water resources and law and independent of the Parties involved in the dispute. It is preferable that the Commissioners adopt panel selection criteria established by recognized national arbitral organizations. An optional framework for the Arbitral Panel is presented at Annex C-2.

A panel of five-member panel is recommended, although that number is somewhat arbitary. An odd number was chosen to assure a decision can be reached by majority vote. The use of a single panel member is discouraged since it is unlikely that all Parties could agree on the selection. Also a decision by a single arbitrator is usually difficult to accept by all Parties, especially those for whom the decision is negative. The number five has been chosen as a means of reducing a conscious or unconscious bias that may effect the decision.

A major consideration of the Parties is whether the decision of the Arbitral Panel will be final or to allow further conflict resolution by litigation. On the one hand, Arbitration finality allows for faster and more equitable conflict resolution; if the framework of the Panel is properly established, Arbitration finality provides for conclusive resolution by persons acknowledged as experts in the field. However, the Parties may consider that the use of Arbitration finality lacks sufficient judicial competence for such judgments to be politically acceptable in their particular domestic circumstances. If the decision is made to include ADR with binding arbitration in the Agreement, §6C-1-07(d) should be included and §6C-1-08 (Right to Litigate) omitted. If the Parties choose to include ADR in the Agreement but do not choose to include binding arbitration, §6C-1-07(d) should be omitted and §6C-1-08 (Right to Litigate) included.

Cross-references: §1C-1-01 (general policies); §1C-1-02 (purposes of agreement); §1C-1-03 (objectives of agreement); §1C-1-04 (coordination and cooperation); §2C-1-02 (consent to jurisdiction); §2C-1-03 (duration of agreement); §2C-1-04 (amendments and supplements); §2C-1-05 (powers of sovereign parties; withdrawal); §2C-2-09 (party or parties); §4C-1-01 (joint exercise of sovereignty).

Similar Agreements: *Convention on the Law of the Sea*, 33 I.L.M. 1309 (1982); *Convention on the Protection and Use of Transboundary Watercourses and International Lakes*, 31 I.L.M. 1312 (1992); *North American Free Trade Agreement*, 19 U.S.C. §§ 3311-3473 (1993); *Treaty of Peace between the State of Israel and the Hashemite Kingdom of Jordan*, 34 I.L.M. 43 (1994); *Delaware River Basin Compact*, Pub. L. 87-328, 75 Stat. 688 (1961); *Susquehanna River Basin Compact*, Pub. L. No. 91-575, 84 Stat. 1509 (1970).

§6-1-08 RIGHT TO LITIGATE (Optional)

Nothing in this agreement shall be construed to limit or prevent either Party from instituting or maintaining any action or proceeding, legal or equitable, in any tribunal of competent jurisdiction for the protection of any right under this agreement or the enforcement of any of its provisions.

Commentary: If the Parties choose not to adopt a provision which states that the decision of the Arbitral Panel shall be considered final and shall not be appealable to any court of law or equity [*see* § 6C-1-04], the Agreement must provide for appropriate legal or equitable action or proceeding. The existence of an appropriate tribunal may pose a problem in cases not involving an entity like the United States or European Union. It may be advisable to specify the tribunal in the agreement itself to avoid dispute over jurisdictional questions at a later date.

Cross References: §1C-1-01 (general policies); §1C-1-02 (purposes of agreement); §1C-1-03 (objectives of agreement); §1C-1-04 (coordination and cooperation); §2C-1-02 (consent to jurisdiction); §2C-1-03 (duration of agreement); §2C-1-04 (amendments and supplements);

§2C-1-05 (powers of sovereign parties; withdrawal); §2C-2-09 (party or parties); §4C-1-01 (joint exercise of sovereignty).

Similar Agreements: *Belle Fourche River Compact*, 58 Stat. 94 (1944); *Colorado River Compact*, 45 Stat. 1057 (1928); *Snake River Compact*, 64 Stat. 29 (1949).

SIGNATURES

IN WITNESS WHEREOF, and in evidence of the adoption and enactment into law of this Agreement by the signatory Parties, the representatives of the sovereign States of _____. _____. _____ do hereby, in accordance with authority conferred by law, sign this Agreement in (six) duplicate original copies, as attested by the appropriate authorities of the respective sovereign Parties, and have caused the seals of the respective States to be hereunto affixed this ____ day of _____.

ANNEX C-1

FINANCING

(Example)

1. ANNUAL CURRENT EXPENSE AND CAPITAL BUDGETS.

(a) The Commission shall annually adopt a capital budget that includes any and all capital projects it proposes to undertake or continue during the budget period. This budget shall contain a statement of the estimated cost of each project and the method of financing thereof. Revenues accruing to the capital projects will be credited to the Commission's capital account, and will be used to defray annualized capital expenses.

(b) The Commission shall annually adopt a current expense budget for each fiscal year. Such budget shall include the Commission's estimated expenses for administration, operation, maintenance and repairs. It shall include a separate statement for each project, together with its cost allocation. The total of such expenses shall be balanced by the Commission's estimated revenues from all sources, including the cost allocations undertaken by any of the signatory Parties in connection with any project.

(c) The Parties agree to include the amounts so apportioned for the support of the current expense budget in their respective budgets next to be adopted, subject to such review and approval as may be required by their respective budgetary processes. Such amounts shall be due and payable to the Commission in quarterly installments during its fiscal year, provided that the Commission may draw upon its working capital to finance its current expense budget pending remittances by the signatory Parties.

2. CAPITAL FINANCING BY SIGNATORY PARTIES; GUARANTEES

The Parties will provide such capital funds required for projects of the Commission as may be authorized by their respective statutes in accordance with a cost sharing plan prepared by the Commissioners; but nothing in this section shall be deemed to impose any mandatory obligation on any of the Parties other than such obligations as may be assumed by a Party in connection with a specific project or facility.

3. GRANTS, LOANS OR PAYMENTS BY PARTIES

(a) The Commission may receive and accept, and the Parties may make, loans, grants, appropriations, advances and payments of reimbursable or non-reimbursable funds or property in any form for the capital or operating purposes of the Commission.

(b) Any funds which may be loaned to the Commission either by a Party or a political subdivision thereof shall be repaid by the Commission through the issuance of bonds or out of other income of the Commission, such repayment to be made within such period and upon such terms as may be agreed upon between the Commission and the signatory Party or political subdivision making the loan.

4. RATES AND CHARGES

The Commission may from time to time after public notice and hearing fix, alter and revise rates, rentals, charges and tolls and classifications thereof, for the use of facilities which it may own or operate and for products and services rendered thereby, without regulation or control by any department, office or agency of any signatory Party.

5. BORROWING POWER

(a) The Commission may borrow money for any of the purposes of this Agreement, and may issue its negotiable bonds and other evidences of indebtedness in respect thereto.

(b) All such bonds and evidences of indebtedness shall be payable solely out of the properties and revenues of the Commission without recourse to taxation. The bonds and other obligations of the Commission, except as may be otherwise provided in the indenture under which they were issued, shall be direct and general obligations of the Commission and the full faith and credit of the Commission are hereby pledged for the prompt payment of the debt service thereon and for the fulfillment of all other undertakings of the Commission assumed by it to or for the benefit of the holders thereof.

6. CREDIT EXCLUDED

The Commission shall have no power to pledge the credit of any signatory Party or to impose any obligation for payment of the bonds upon any Party or sub-division thereof. Neither the Commission nor any person executing the bonds shall be liable personally on the bonds of the Commission or be subject to any personal liability or accountability by reason of the issuance thereof.

7. BONDS; AUTHORIZATION GENERALLY

(a) Bonds and other indebtedness of the Commission shall be authorized by resolution of the Commissioners.

(b) The Commission may issue bonds in one or more series and may provide for one or more consolidated bond issues, in such principal amounts and with such terms and provisions as the Commission may deem necessary. The bonds may be secured by a pledge of all or any part of the property, revenues and franchises under its control.

(c) Bonds may be issued by the Commission in such amount, with such maturities and in such denominations and form or forms, whether coupon or registered, as to both principal and interest, as may be determined by the Commission.

(d) The Commission may provide for redemption of bonds prior to maturity on such notice and at such time or times and with such redemption provisions, including premiums, as the Commission may determine.

9. FUNDING AND REFUNDING

Whenever the Commission deems it necessary, it may fund and refund its bonds and other obligations, whether or not such bonds and obligations have matured.

10. REMEDIES FOR DEFAULT ON BONDS (Optional)

(a) The holder of any bond may for the equal benefit and protection of all holders of bonds similarly situated require and compel the performance of any of the duties imposed upon the Commission or assumed by it, its officers, agents or employees under the provisions of any indenture, in connection with the acquisition, construction, operation, maintenance, repair, reconstruction or insurance of the facilities, or in connection with the collection, deposit, investment, application and disbursement of the rates, rents, tolls, fees, charges and other revenues derived from the operation and use of the facilities, or in connection with the deposit, investment and disbursement of the proceeds received from the sale of bonds.

(b) In the alternative, the holder of any bond may, by action or suit in a court of competent jurisdiction of any signatory Party, require the Commission to account as if it were the trustee of an express trust, or enjoin any acts or things which may be unlawful or in violation of the rights of the holders of the bonds.

(c) The enumeration of such rights and remedies does not, however, exclude the exercise or prosecution of any other rights or remedies available to the holders of bonds.

11. ANNUAL INDEPENDENT AUDIT

(a) A yearly audit shall be made of the financial accounts of the Commission. The audit shall be made by qualified certified public accountants, selected by the Commissioners, who have no personal interest direct or indirect in the financial affairs of the Commission or any of its officers or employees. The report of audit shall be prepared in accordance with accepted accounting practices and shall be filed with the chairman and such other officers as the Commission shall direct. Copies of the report shall be made available for public distribution.

(b) Each Party shall be entitled to examine and audit at any time all of the books, documents, records, files and accounts and all other papers, things or property of the Commission. The representatives of the Parties shall have access to all books, documents, records, accounts, reports, files and all other papers, things or property belonging to or in use by the Commission and necessary to facilitate the audit and they shall be afforded full facilities for verifying transactions with the balances or securities held by depositaries, fiscal agents and custodians.

ANNEX C-2

ALTERNATIVE DISPUTE RESOLUTION

ANNEX C-2

ARBITRAL PANEL
(Optional)

1. ROSTER AND QUALIFICATIONS OF PANELISTS

(a) The Commission shall establish and maintain a roster of up to 30 individuals who are willing and able to serve as panelists. The roster membership shall be approved by the Parties to this Agreement. The roster members shall be appointed by consensus for terms of three years, and may be reappointed.

(b) Roster members shall have expertise or experience in law and in water resources engineering or related discipline, environmental engineering or related ecological discipline or water resources economics, or other matters covered by this Agreement and shall be chosen strictly on the basis of objectivity, reliability and sound judgment;

(c) Panel members shall comply with a code of conduct to be established by the Commission.

2. PANEL SELECTION

(a) Where there are two disputing Parties, the following procedures shall apply: (i) The panel shall comprise five members. (ii) The disputing Parties shall endeavor to agree on the chair of the panel within 15 days of the delivery of the request for the establishment of the panel. If the disputing Parties are unable to agree on the chair within this period, the disputing Party chosen by lot shall select within five days as chair an individual who is not a citizen of that Party. (iii) Each disputing Party, within 15 days of selection of the chair, shall select two panelists who are citizens of the other disputing Party. (iv) If a disputing Party fails to select its panelists within such period, such panelists shall be selected by lot from among the roster members who are citizens of the other disputing Party.

(b) Where there are more than two disputing Parties, the following procedures shall apply: (i) The panel shall comprise five members. (ii) The disputing Parties shall endeavor to agree on the chair of the panel within 15 days of the delivery of the request for the establishment of the panel. If the disputing Parties are unable to agree on the chair within this period, the Party or Parties on the side of the dispute chosen by lot shall select within 10 days a chair who is not a citizen of such Party or Parties. (iii) Within 15 days of selection of the chair, the Party complained against shall select two panelists, one of whom is a citizen of a complaining Party, and the other of whom is a citizen of another complaining Party. The complaining Parties shall select two panelists who are citizens of the Party complained against. (iv) if any disputing Party fails to select a panelist within such period, such panelist shall be selected by lot in accordance with the citizenship criteria of subparagraph (3).

(c) Panelists shall normally be selected from the roster. Any disputing Party may exercise a peremptory challenge against any individual not on the roster who is proposed as a panelist by a disputing Party within 15 days after the individual has been proposed.

(d) If a disputing Party believes that a panelist is in violation of the Model Code of Conduct for Arbitration, the disputing Parties shall consult and if they agree, the panelist shall be removed and a new panelist shall be selected in accordance with this Section.

3. MODEL RULES OF ARBITRAL PROCEDURE

(a) The Commission shall establish within one year of the entry into force of this Agreement Model Rules of Procedure, in accordance with the following principles: (i) the procedures shall assure a right to at least one hearing before the panel as well as the opportunity to provide initial and rebuttal written submissions; and (ii) the panel's hearings, deliberations and initial report, and all written submissions to and communications with the panel shall be open to the public.

(b) Unless the disputing Parties otherwise agree, the panel shall conduct its proceedings in accordance with the Model Rules of Arbitral Procedure.

(c) Unless the disputing Parties otherwise agree within 20 days from the date of the delivery of the request for the establishment of the panel, the terms of reference shall be:

"To examine, in the light of the relevant provisions of the Agreement, the matter referred to the Commission (as set out in the request for a Commission meeting) and to make findings, determinations and recommendations."

4. PANEL DECISION; INITIAL REPORT

(a) Unless the disputing Parties otherwise agree, the panel shall base its report on the submissions and arguments of the Parties and its independent analysis of the dispute.

(b) Unless the disputing Parties otherwise agree the panel shall, within (____) days after the last panelist is selected, present to the disputing Parties its recommendations, if any, for resolution of the dispute.

(c) Panelists may furnish separate opinions regarding the initial report on matters not unanimously agreed.

(d) A disputing Party may submit written comments to the panel regarding the initial report within 14 days of presentation of the report.

(e) In such an event, and after considering such written comments, the panel, on its own initiative or on the request of any disputing Party, may: (i) request the views of any participating Party; (ii) reconsider its report; and (iii) make any further examination that it considers appropriate.

5. PANEL DECISION; FINAL REPORT

(a) The panel shall present to the disputing Parties a final report, including any separate opinions on matters not unanimously agreed, within 30 days of presentation of the initial report, unless the disputing Parties otherwise agree.

(b) No panel may, either in its initial report or its final report, disclose which panelists are associated with majority or minority opinions.

(c) The disputing Parties shall transmit to the Commission the final report of the panel, including the independent analysis of the panel, as well as any written views that a disputing Party desires to be appended, within a reasonable period of time after it is presented to them.

(d) Unless the Commission decides otherwise, the final report of the panel shall be published 15 days after it is transmitted to the Commission.

6. IMPLEMENTATION OF FINAL REPORT

(a) On receipt of the final report of a panel, the disputing Parties shall agree on the resolution of the dispute, which normally shall conform with the determinations and recommendations of the panel, and shall notify the Commission of any agreed resolution of any dispute.

(b) There shall be no appeal from the determinations in the final report.

Index

activities: prohibited 120–121; within territory of other party 58–61
administration 16, 19–32, 51, 100, 107–129; agreement 56–61; by commission 21–28; of commission 31–32, 113–114; officials 56–58
administrative authority 19–28, 107–114
administrative support 20, 57
advisory boards 27–28
advisory committees 122
agencies, existing 15–16, 99–100
agreement: administration 56–61; duration of 13–14, 49, 97–98; geographic scope 17; implementation and verification 20, 57–58; material breach 166–167; modification 165–166; negotiations and consultations 167–168; objectives 90–91; purposes 42–43, 88–90; scope 42–43, 88–90; termination 13–14, 49; voiding 16, 50–51
Agreement for Coordination and Cooperation in the Management of Shared Water Resources 4–37
agricultural practices 35, 68, 158
allocation, water 33–34, 42–43, 63, 115, 132, 142–151; alternatives 72–78; criteria 148–150; during drought 140–141; economic evaluation 149–150; existing uses 72–73; during floods 138–139; formulas 153–154; guaranteed minimums 75–76; guaranteed quantity 76–77; methodology 148; by percentages 72–73; prioritization 142; proportionate 73–75; specific purposes 41; storage 78; surface models 65; vested rights protected 76–77; waters not subject to 143–144
amendments to agreements 15; process 50, 99
annexes 16, 51, 100
applicability, limited 50–51
appropriative rights 63, 123–124
aquatic life 91
aquifers 17, 52, 72, 101, 106, 109, 145
arbitration 171–172
artesian flow 55
atmospheric water 43, 55–56, 73, 89, 106, 160–161; defined 18–19, 52, 101
audits 176–177
authorities: coordination of 20; geographic 23; hydrologic 23

basin: assessment of water demands 145–148; classification of water demands 145–148;
conceptual model vii; defined 17, 52, 101; hydrologic mass balance vii; water sources assessment 144–145; water sources enhancement 144–145; waters 55–56; waters, defined 18–19, 106–107
Bell Fourche Compact 12, 48, 73, 96
Big Blue River Compact 65, 159
bonds 175–176; default remedies 176
bribes 120
budgets 115, 117, 162, 174

Canadian River Compact 78
canals 59
capillary moisture 55, 106
capital costs 14, 49, 97, 98
catchment areas 35, 68, 158; management approach 68, 158
claims, filing 59
climate change 49, 136–138
climatology 87
Colorado River Compact 76
comity: international 42–43, 88; interstate 42–43, 88
commission: administration by 21–28; administration of 113–114; authorities 23; created 21–22, 107–108; cultural differences of parties 108; funding and expenses 32; immunities 25–26, 110–112; jurisdiction 108–109; legal protections 25–26, 110–112; naming 21–22; official communications 25, 111; organization 26–27, 112–113; powers and duties 28–31, 115–126, 118–119; privileges 25–26, 110–112; property and assets 25, 111; responsibilities 23; staffing 26–27, 112–113; status 25–26, 110–112
commissioners 23–25, 109–110; administration 116–118; international 24; powers and duties 116–118; for two-party agreements 24, 109–110
comprehensive water management 130–162
Comprehensive Water Management Agreement 84–180
comprehensive water management plan 131–133; conditions 135–136; defined 102; deviation from 136–138; impairment 141; purpose and objectives 133–135
condemnation 59–60, 123–124. see also takings
conduits 59
conflict resolution 23, 33, 36–37, 41, 69–70,

181

107, 163–173. *see also* dispute resolution; arbitral panel 178–180; arbitration 171–172; conciliation and mediation 170–171; judicial 169; negotiations and consultations 167–168; by signatory parties 36–37, 70
Congressional approval 10, 12, 13, 46, 48, 94, 95–97; amendments 50, 99
conservation measures 52, 58, 102–103, 140; defined 52–53, 102–103; examples 53
consumption: defined 53–54; water 33–34
consumptive water use 34–36, 52, 102–103, 134, 145
contractual obligations 9
control, exclusive 33–34, 62–63
Convention on the Law of the Non-Navigational Uses of International Watercourses 4, 54, 68, 84, 87, 89, 104, 132, 142–143, 157–158, 164
cooperation 8–9, 31–32, 43–45, 91–93, 126–127; agency 21–22; cooperation/coordination model 4; governmental 21–22; intergovernmental 86–87
cooperative services 129
coordination. *see* cooperation
corruption 120–121

damage liability 45
dams 17, 34–36, 64, 78; hydropower 35
data collection 15, 23, 115
data dissemination 23
data exchange 30–31, 34–36. *see also* information exchange; meetings 34–36
definitions 17–19, 52–56, 101–107
deforestation 35, 68, 158
Delaware River Basin Compact 24, 84, 110, 119, 120, 163
demands, water vii; baseline 132
development: sustainable 103; water resources 34–36
development, water 34–36
dispute resolution 36–37, 41, 69–70, 107, 163–173. *see also* conflict resolution; alternatives 168–170, 178–180; arbitral panel 178–180; arbitration 171–172; conciliation and mediation 170–171; judicial 169; negotiations and consultations 167–168; by signatory parties 70
diversion, water 33–34, 59, 62, 64, 73, 76–77, 145, 154–155; defined 53–54; exclusive control 33–34
drainage 17
drought 74, 134–138; allocation during 140–141; defined 17, 53, 103; management strategy 53, 88, 103

ecology 91, 115
economic analysis 134
economic growth 135–136; impairment 35, 68, 157
effective date of commission agreement 11–13, 47–49, 95–96
effluent 35, 67, 68–69, 157, 158
emergencies, water 22
eminent domain 59–60, 123–124. *see also* takings
enforcement 119–120
environmental protection 133, 135–136
equitable apportionment 54, 104
equitable funding 32
expenses, commissions 32
experts 27, 122

federal agencies 95
federal rights 10, 46–47, 93–94
financing 21, 49, 58, 162–163, 174–177; capital 174
floods 134–138; allocation during 138–139; damage 64; defined 17–18, 54, 104–105; flood plains 139; protection programs 34–36, 65–66, 88, 152–153
flow: criteria 67, 157; measurement 63; minimum 153–154; reduction 64–65; surface 65
flushing 69
freshwater systems 68, 158
funding 21, 58; commissions 32; international 32

gaging stations 20, 63, 73, 74–77
gifts 120
good faith implementation 9–10, 45–46, 164–165
government reorganization 20
groundwater 55, 106

habitats 151
health, public 133, 135–136
health risks 35, 68, 157
hearings: commissions 31–32, 125–126; public 125, 135–136
hydrologic cycle 34–36
hydrologic mass balance vii
hydrologic record 14, 49, 97
hydrology 49
hydropower 90–91; dams 35; facilities 64; sales of facilities 125

importation, water 66–67, 73, 107, 155–156
industrial plants 35, 68, 158

industrial waste 35, 68, 158
information dissemination 115–116
information exchange 30–31, 34–36, 44, 92.
    see also data exchange; meetings 34–36
inter-state provisions 48
interbasin transfer 18, 34–36; defined 18, 105; surface water 141
intergovernmental relations 126–129
international-related clauses 11, 14–15, 24, 25, 47, 61, 68–69, 94, 95–96, 98, 111
intrabasin transfer 141
involvement, public 115
irrigation 72, 90–91; drip 53, 103; wells 159
issues, water 33–36

joint expenses 112
jurisdiction: consent to 13, 49–50, 96–97; exclusive 33–34, 62–63
just compensation 60, 123–124

La Plata Compact 75
lakes 55
legal protections, commission 25–26, 110–112
levels, water 64–65
Limited Purpose Agreement for the Shared Use of Transboundary Water Resources 41–78
litigation 13, 97
litigation rights 37, 70, 168, 172–173

management, water 87; officials 19–20, 56–57; regulations 120
maps 17, 52, 101
market mechanisms 134
McCarren Amendment 13, 97
meetings: commission 31–32, 125–126; public 117
Model Code for the Shared Use of Transboundary Water Resources iii

national security 11, 47, 94
natural resources 91
navigational interests 61
non-consumptive water use 134

obligations, general 47–51, 95–100
officials 19–21; administration 56–58; party administration 19–21; state 95; substitution 20, 57; water management 56–57
optional clauses: advisory committees 122; agreement duration 13–14, 49, 97–98; agreement objectives 90–91; amendments and supplements 15, 50, 99; approval of water resource projects 121–122; arbitration 178–180; augmentation of supply 66–67; commission privileges and immunities 111; commissions 25, 27–28, 30–31; comprehensive water management plan 135–136; condemnation 123–124; consent to jurisdiction 13; dispute resolution 69–70; eminent domain 59–60; existing government agencies 15–16; federal rights 10; flood protection programs 65–66; information exchange 34–36; limited applicability 50–51; litigation rights 172–173; meetings, hearings, and records 125–126; national security 11, 47; powers and duties of the commission 118–119; powers and duties of the commissioners 117; powers of sovereign parties 14, 98; preservation of federal rights 46–47; recreational use 161–162; reports 122–123; rights in territory of other party 58–59; storage and diversion 59; underground water 65; underground water withdrawals 159–160; water allocation 63, 148–150; water levels protected 64–65; water quality 67–69; watershed management 150–151; withdrawals and diversions 154–155

parties, defined 18, 55, 105–106
planning 87
planning horizon 134
policies: development 21–22, 107–108, 117; general 6–8, 86–87; planning 44; and purposes 42–47, 86–94
pollution 67–69, 87, 157
powers and duties: commissions 28–31, 115–126; general 28–30, 115–116; special 30–31
precipitation augmentation 66–67, 160–161
private water rights holders 130
procedures 16, 51, 100; commissions 154–155; rules of procedure 28, 113
projects, water: costs 127–128; evaluation criteria 127–128; management 44; planning 44, 128–129
protected areas 154–155
provisions, general 11–19, 47–56, 95–107, 152–162

quality, water 34–36, 67–69, 87, 156–159; control 41; criteria 67, 69, 156; degradation 35, 68, 158; maintenance 88–89; management 132; measurement 20
quantity, water 20

receiving waters 67
recharge capability 109
recharge rate 145

records: commission 31–32, 125–126; public 31–32, 125
recreational use 34–36, 90–91, 133, 161–162
Red River Compact 13, 50, 63, 97
regulation, water 58
regulations, agreement 119–120
rejuvenation, water 69
reports 122–123
Republican River Compact 12, 48, 96
reservoirs 17, 34–36, 55, 64, 72–73; storage 59; water supply 64
resources, water 7; allocation 62–69; assessment vii, 134; database vii; development 42–43, 49, 97; geographic scope 87; integrated management 134; interrelationship 131; management 15, 89, 90–91; projects 121–122
return flows 53, 102–103
rights: riparian water 63; within territory of other party 58–59; water 152
Rio Grande Compact 13, 49–50, 73, 97
riparian lands 67, 68, 84, 151, 157
rivers 55
rules and regulations 119–120
rules of procedure 28, 113
runoff 151

Sabine River Compact 63
severability of agreement 16, 100
sewage treatment 35, 68, 158
shared water agreements, legal evolution ix–xiv
sharing, water 41
shortages, water 7, 107
showerheads, low flow 53, 103
signatory parties, projects 128–129
siltation 158
soil erosion 151, 158
soil moisture 55, 106
sources, water vii
South Platte River Compact 77
sovereign equality 8, 44, 91
sovereign immunity 97
sovereign parties 14–15, 98
sovereign rights 4, 33
sovereignty 36, 98, 116; joint exercise 130–131; relinquishment 86–87
springs 55
storage, water 59
stream adjudication, general 97
streams 55
supplements to agreement 15; process 50, 99
supply augmentation 34–36, 66–67, 155–156
supply, public 77
surface models, guide to allocations 65

surface water 35, 55–56, 68, 87, 89, 106, 109, 130, 134; assessment 144–145; basins 52, 101; defined 18–19; degradation 158; diversion 154; withdrawal 154
Susquehanna River Basin Compact 24, 84, 110, 120
sustainability 35, 68, 157
sustainable development 90–91, 103
SUTWR. *see* Model Code for the Shared Use of Transboundary Water Resources

takings 48, 59–60, 96, 123–124
taxation 10, 46, 60, 93–94, 111
technical support 20, 57
territorial integrity 8, 44, 91
Texas v. New Mexico 9, 13, 45, 50, 78, 97
toilets, low flow 53, 103
tolls 61
tort 62; liability 126
Treaty between the United States and Great Britain relating to Boundary Waters 4
tribunals 37, 50–51, 172
tributaries 17, 52, 101

underground water 35, 43, 55–56, 68, 73, 87, 89, 106, 109, 130, 134; assessment 145; defined 18–19, 55, 106; degradation 158; diversion 154; limits on withdrawals 159–160; withdrawal 65, 154
United States-related clauses 10, 12, 13, 48, 49–50, 67–68, 93–97
urbanization 35
use, water 115; categories 146–148
utilization, water 33–36, 90–91; defined 53–54, 104; exclusive control 33–34

vested water rights 143–144

water allocation. *see* allocation, water
water, as a commodity 149
water, atmospheric. *see* atmospheric water
water, potable 87
water-saving strategies. *see* conservation measures
water, underground. *see* underground water
water value definition 149
watersheds 17, 52, 84, 101; management 150–151
wells 55, 65; irrigation 159
wetlands 67, 157
withdrawal from agreements 14–15, 98
withdrawal, water 52, 102–103, 154–155; underground water 159–160